The Urban Book Series

Aims and Scope

The Urban Book Series is a resource for urban studies and geography research worldwide. It provides a unique and innovative resource for the latest developments in the field, nurturing a comprehensive and encompassing publication venue for urban studies, urban geography, planning and regional development.

The series publishes peer-reviewed volumes related to urbanization, sustainability, urban environments, sustainable urbanism, governance, globalization, urban and sustainable development, spatial and area studies, urban management, urban infrastructure, urban dynamics, green cities and urban landscapes. It also invites research which documents urbanization processes and urban dynamics on a national, regional and local level, welcoming case studies, as well as comparative and applied research.

The series will appeal to urbanists, geographers, planners, engineers, architects, policy makers, and to all of those interested in a wide-ranging overview of contemporary urban studies and innovations in the field. It accepts monographs, edited volumes and textbooks.

More information about this series at http://www.springer.com/series/14773

Agata Bonenberg

Cityscape in the Era of Information and Communication Technologies

Springer

Agata Bonenberg
Poznań University of Technology
Poznań
Poland

ISSN 2365-757X ISSN 2365-7588 (electronic)
The Urban Book Series
ISBN 978-3-319-69541-9 ISBN 978-3-319-69542-6 (eBook)
https://doi.org/10.1007/978-3-319-69542-6

Library of Congress Control Number: 2017955666

Printed on acid-free paper

This Springer imprint is published by Springer Nature
The registered company is Springer International Publishing AG
The registered company address is: Gewerbestrasse 11, 6330 Cham, Switzerland

Contents

1 Introduction 1
References 8

2 Architectural Concept: Location Versus Tradition of a Design Philosophy 9
References 21

3 Mass Communication and Public Space 23
3.1 Elite Stage—Compact and Clear City Composition 38
3.2 Popularization Stage—Polycentric Composition 41
3.3 Specialization Stage in the Industrial Cities of the Twentieth Century—Specialized System Versus Loss of Legibility of Urban Structures 42
3.4 Multimedia Stage of Mass Communication—Urban Spaces of Information Society 44
References 48

4 Strategies for the Development of Public Spaces in the Era of Mass Communication 49
4.1 Social Media 51
4.2 Public Consultation 54
4.3 Urban Space Activation 54
4.4 Design Verification 55
References 56

5 Cityscape: Dominant Functions and the Role in the City Structure 57

6 Designing the Living and Working Space of the Creative Class . . . 65
6.1 The Creative Process and Spatial Environment 66

6.2 In Search for Identity: Advanced Technologies, Natural and Historical Values in Architecture for Creative Class 72
References 74

7 Architectural Landscape in Mass Media Coverage 75
7.1 Architecture as Scenography in the Mass Media Event, Supporting Creation of Place—Brand 75
7.2 New Stylistic Architectural Forms Based on the Creation and Distribution of Symbols and Metaphors 78
7.3 Place as the Subject of Fashion Created by the Media 85
7.4 The Perception of Architecture on Online Maps—The Fifth Facade 96
7.5 Media Message as an Element of Architectural Space 96
References 101

8 Place Brand-Building: Influencing Emotional Perception of Urban Spaces Through Social Media 103
8.1 Mapping Emotions 105
8.2 Sense of Community 106
References 107

9 Spatial Transformations and Work-Related Mobility 109
9.1 Expansion: Increase in Use of Space 109
9.2 Inversion: The Loss of Function 113
9.3 Intensification of Development and Transformations Within the Existing Structure of the City 113
9.4 Adaptation of Existing Structures 115
9.5 Reconfiguration—Kinetic Architecture 119
9.6 Negotiation—Changes of a Functional Space Division 124
References 132

10 New Urban Developments in Genoa, Milan and Trento as an Illustration of Spatial Transformations Designed in Response to Change in Socio-economic System 133
10.1 Genoa, the “Waterfront” Project 135
10.2 Milan, the “FALCK” Project 148
10.3 Trento, Project for the Post-industrial Areas of Michelin 149
10.4 Building an Image Based on the Relationship with the Existing Context of the City and Development of the Social Communication Space 151
References 152

11 Conclusion 153

Chapter 1
Introduction

Abstract The urban structure of a city and its architecture are the spatial manifestation of the culture of its inhabitants. Historic urban spaces provide a basis for observation and drawing conclusions regarding the past social, economic, cultural relations. An evolutionary order of creating new typologies of space and construction technologies is emerging from the research on the history of architecture in response to the demand and new functions. The ornament and aesthetics following the cultural changes of society are subject to evolutionary changes.

The urban structure of a city and its architecture are the spatial manifestation of the culture of its inhabitants. Historic urban spaces provide a basis for observation and drawing conclusions regarding the past social, economic, cultural relations. An evolutionary order of creating new typologies of space and construction technologies is emerging from the research on the history of architecture in response to the demand and new functions. The ornament and aesthetics following the cultural changes of society are subject to evolutionary changes.

The motivation to take up the topic is the desire of referring the relations between the community and the built environment, so perfectly studied and described in many historical examples, to the present day conditions. Interest in issues of directionality and purpose in designing architecture of the "here and now", realized in practice when working with the architect Renzo Piano, prompted the author to search for sources of contemporary concepts of spatial design. This common source is the basic factor shaping the contemporary culture—a wide access to information.

The main goal of this work is, therefore, to determine the influence of the contemporary media (means of mass communications) on the trends shaping the architectural space. This influence can be seen on many levels: from changes in settlement preferences, due to changes in the formula of employment, to the question of the mass media promoting trends in stylistics, triggering the need for mass customization of working and living space. It is important to emphasize the applicative value of the research and cited solutions verified with the experience of the author, acquired while working in architectural offices in Germany, Australia, Scotland and Poland. An important experience that has shaped the views of the author on the discussed topics was the collaboration with the Architectural Association School of Architecture in London (AA) and participation in the

A. Bonenberg, *Cityscape in the Era of Information and Communication Technologies*, The Urban Book Series,
https://doi.org/10.1007/978-3-319-69542-6_1

seminar on *social media* at Harvard University in Cambridge, USA. At each of the mentioned stages of work, it proved important to take into consideration the changes to which the community is subject through the new media. The teaching positions at Politecnico di Milano and Poznan Univesity of Technology gave the author the opportunity to research and study the subject in the didactic process.

Information technology plays an important role among the contemporary mass media. Building of an information society is one of the objectives of a social policy in Poland and the European Union. A historical report "Europe and the Global Information Society—Recommendations to the European Council" elaborated in 1994 by the Bangemann's Commission,[1] pointed directions and proposed rules for the organization of the European information society. It describes the opportunities arising from the possibilities of economic development while warning against omissions in this sphere. Omissions and lack of computerization will result in underdevelopment and marginalization of some regions of the country, and its residents can face a technological exclusion, disastrous on a common, competitive job market: *The first countries to enter the information society will gain the most benefit. They will set the agenda for all those, who will follow in their footsteps.*[3] Poland entered the nineties as a pre-information society. In the strategy of the country development, on the basis of the material produced during the works of the First Congress of Polish Information Technology in Poznan in 1994, it was determined that one of the main factors of economic development should be precisely computerization. Cultural transformation caused by the universal access to media using the latest information and communication technologies are not without effect on the architectural space and built environment. According to leading sociologists, urban planners and architects, technological and social changes involve spatial changes. They will cover planning at the regional level, city level as well as architectural design, to the designs of interior furnishings. The thematic selection allowed to expose trends which, according to the author, have the most important impact on the image of the contemporary spatial reality subject to the influence of media. It was important to synthesize the issues to be able to see a coherent logic of the projects implemented today in the field of architecture and urbanism.

From the ancient times, the public spaces in cities served to exchange information—they were designed to promote the flow of information process and the need for social interaction was one of the factors shaping the spatial planning of the urban structures. An important issue raised in the paper is the impact of media coverage on the clarity of the functional structure in the city scale in conjunction with the risk of generating a space devoid of identity.

Reflections on the media, architecture and their relationship to space are initiated by the author by defining the meaning that she desires to associate to these notions. This is a difficult task because of the ambiguity of the semantic scope of the

---

[1]The Commission deliberating under the supervision of Martin Bangemann, the commissioner of the European Union in the years 1993–1999, responsible for the development of telecommunications and information technology—the main promoter of the concept of "information society" in Europe.

categories. They involve different scales, different fields of human activity. Over the centuries, this range was subject to changes. In the colloquial sense, architecture is the art of shaping the space, the science of designing and implementing buildings and their equipment. It also includes designing of a settlement network, cities, towns, villages, settlements and adapting the landscape to human needs. Such a wide scope of understanding of architecture is probably the most similar to the definition coined by Vitruvius, who wrote that it is an extremely extensive "branch of knowledge" which covers "information from various sciences". Here, he pointed out the technical expertise, geometry, drawing skills, optics, arithmetic, astronomy, philosophy, law, knowledge of natural sciences. According to Vitruvius, the architect should be familiar with sculpture, painting, music, master the art of grammar and rhetoric.

The term "media" has a much shorter history. According to the "Oxford English Dictionary"—it came into circulation in the twenties of the twentieth century. This does not mean that the knowledge of communication had not existed much earlier. In ancient Greece and Rome, the art of persuasion, rhetoric and grammar skills were considered to be the basis of education. In the Middle Ages, rhetoric with dialectic and grammar formed a *trivium* of liberal arts (Grant 1996). The art of rhetoric enjoyed a high reputation in the Renaissance (Lichański 2000). In the eighteenth century, rhetoric still played a significant role, with the emphasis put on the style of speech, filled with complex figures of speech, having more reference to emotion than the logic of conveyed message. During this time was developed the term "public opinion", and in the early nineteenth century, the press—a new medium shaping the views—started to play an important role in politics, culture and social life. The twentieth century greatly expanded the use of the press as a means of shaping the public opinion. The outbreak of the First World War directed media attention to a propaganda, involving the intellectual and emotional persuasion aimed at shaping certain behaviours and attitudes of individuals and communities.

Chomsky (2002) ascribes the effectiveness of propaganda to the support of the public authorities and the educated elite and the impossibility of real criticism against the content of propaganda. According to the author, it is mostly achieved through censorship, conspiracy of silence in the media or joint action of publishers.

In 1944, for the first time appeared the term of "cultural industry". Horkheimer and Adorno (1979) used this term in reference to cultural goods produced in order to make a profit. They pointed out that the aim of their development is not to create aesthetic values but profit, and methods of producing cultural goods are no different from industrial production. Therefore, the mass-produced newspapers, books, recordings, films require adequate manufacturing and distribution strategies tailored to the target group of customers. They aptly noted that the commercialization of culture led to a "commodity fetishism", which is manifested by the fact that a work of art has for a mass audience only the value he or she paid for its purchase (Horkheimer 1979).

Media are linked to mass communication. In this perspective, the concept of social communication has been developed by Lévi-Strauss (1993), who drew attention to the phenomenon of social exchange of information, the importance of myth and symbol in the shaping of a social identity (Lévi-Strauss 1993). The myth and symbol as part of mass communication are the carrier of social values; it shapes certain behaviours and relationships with the environment (May 1997). The timeless role of myth, symbol and ritual present in mass communication is perceived by Musioł (2012). Luhmann (2007) extended the concept of mass communication with a theory of self-creating, autopoietic systems. These systems are capable of self-recreating allowing the survival and expansion in the process of social communication. Some of these systems may disappear as a result of evolution, others—modify their content (Skąpska 2007).

In connection with such a broad scope of the concept of media and communication, we can ask what is not covered by the issues associated with the media. The approach presented in the study mainly includes ways to deliver and impart information and ideas, form opinions and tastes by the architecture. The scope of analyses includes phenomena associated with the impact of media on architecture, among which are painting, drawing, photography but what also plays an important role is the Internet and modern techniques of multimedia presentations using digital techniques classified as the new media.

Developing the approach of Lévi-Strauss, architecture can be regarded as a carrier of cultural experiences, which is characterized by three important properties.

First of all, architecture as a medium creates an environment that reflects our culture and social experience.

Secondly, architecture is not only meant to meet the most basic needs: protection against climatic conditions, storage of gained food, etc., but it allows us to create ourselves. This allows us to confront the immediate needs with the cultural record. This record is the source of our experience relating to both the specific space (environment), as well as to our ideas about the space. Such an idea we can encode in the form of a building. The building may thus reflect our aspirations, yearnings, whims. In this sense, architecture is not only synonymous with utility, but it also recreates a cultural code and creates reality according to our ideas (imaginary desires). It records our dreams, ambitions, desires. As a result, architecture as a means of social communication enriches not only a space, but also a human.

Thirdly, architecture is inherited. It is the place of our education; we have been facing it since our childhood in a direct form and in mass media (stories, cartoons, movies, etc., which take place against the background of some architectural scenography). No architect creates architecture from scratch by himself or herself, like we do not create any language from scratch. Growing up in a particular linguistic tradition, we learn a language code. This does not mean that we cannot develop the language, introduce new concepts, build new phrases and metaphors. But all of these we are doing in a specific linguistic tradition. The same is true of architecture. Just because someone fights one style, and promotes the other, regardless of their creativity, he or she still uses the same language of architecture as a medium of social communication. Some architectural styles, like some languages, die. Some are struggling for their identity. Others, through the exchange of information in a globalized media reality, become universal. It applies to English language, its geographical (territorial) coverage and its ability to penetrate into other languages, to the same extent as to the "international" style in architecture. In any case, both the language and the architecture are the medium that serves as a means of mass communication.

Architecture involves encoding (recording) the experience of its creator but also finding answers to social needs. Architecture cannot exist in complete isolation from a broadly comprehended environment, that is, from the social environment and the spatial context. The space formed by architecture is a reflection of our history and our present. Using its proper media, it informs about the people who inhabit the area. It allows us to recognize their tastes, the level of culture, economical condition, wealth, habits. It plays the role of media through which we can learn about the surrounding reality.

Interestingly, architecture, like other media, is heavily dependent on political systems. So it was in the distant past, it is also in the recent times. It shall be enough to recall Bauhaus with its modernity, Soviet communism with its socialist realism Mussolini's fascism, National Socialism architecturally interpreted by Albert Speer. In today's reality, the ideology has been replaced by the market. Architecture, like other media, is dependent on the market. To a large extent, it acts as a spatial icon—a kind of medium aimed at attracting capital, it becomes a kind of a spatial brand. Thus, the main research issues of this paper are urban spaces and architecture subject to change under the influence of new mass media. It is the result of a new way of communicating of a society and of new needs arising from it. This applies, at different levels, to places that already exist, are currently reconstructed or built from scratch.

The scientific papers devoted to the discussed topic include many scientific disciplines. This particularly applies to the problems of mass communication, which is the subject of media studies—a relatively new discipline including a broad spectrum of political, economic and cultural issues, associated with the impact of mass communication on the various fields of human activity. In addition to media studies, we can enumerate a number of other disciplines relating to this subject: cultural anthropology, proxemics, social environmental psychology, political science, cultural studies.

Briggs and Burke (2010) in the paper "Social history of media" apply a genetic method, searching for the relevant causal relationships based on analysis of historical texts. Manovich (2011) complements this method with systematics, which defines the basic concepts of new media. Nierenberg (2001) applies the systemic approach to media management with elements of descriptive prediction—predicting of how the media development trends will evolve in future. A similar evolutionary-narrative method is used by Celiński (2010). Noteworthy are elements of the critique of cultural studies in relation to the evolution of media technologies and communication connotations.

Few works in the field of architecture, in principle, do not go beyond these methods; they are complemented with a richer pictorial narrative (reproductions, photographs, sketches). A representative example is the work of Mitchell (1996).

The author to a large extent refers to these research methods. She uses elements of comparative analysis, criticism of literature, diagnosis of causal relationships resulting from the dynamics of the analysed processes. However, she complements these elements with the method of *research by design* (Biggs and Buchler 2008).

This method involves the use of design and artistic practice in order to acquire knowledge. It is based on an occasional experiment in the form of design and architectural drawing. The scientific conclusions lead to systematizing generalizations resulting from the analysis of representative *case studies*. In this perspective, an important research tool is copyright architectural drawings and designs made by the author in the studio Renzo Piano and AA in London. The reputation of these companies–institutions determines the representativeness of the selected examples. The proposed methodology is based on the contextual processing of knowledge. This approach allows the logical fragmentation of the observed dependencies and allows the systematics of the issues included in the distinguished focus groups:

1. Architecture in the age of social communication: location, mobility, architects, architectural education.
2. Cities as an area of mass communication.
3. Social communication versus a city structure.
4. City development strategy in the age of mass communication.
5. Architecture of the information society: the dominant functions and the role in the city structure.
6. Designing personalized space in the context of access to information and communication technologies.
7. Impact of media on architecture.
8. Media release as a part of the architectural space.
9. New language of architecture in the age of social media.

The methodology uses a sequence of relationships between the knowledge discovery and data mining. The rules included in a summary of each chapter make a contextual series of patterns defining the heterogeneous research field.

This research method allows for the integration of large amounts of distributed data belonging to many (seemingly remote) disciplines, such as cultural studies, proxemics, psychology, architecture, theory of architectural design. As a result, the contextual recognition of relationships between media, space and architecture in a dynamically changing environment is characterized by a high degree of credibility.

In relation to the issue of social interaction, an architectural space plays an important role as a place for meeting people, exchanging ideas, information. The communications revolution, which took place at the time of dissemination of information and communication technology, has changed the way of organizing the data flow and social interaction. The emergence of a competitive space for swapping images, information, an alternative space for "meetings" and the mass media impact cannot remain without effect on the spatial expression of human life environment. In this work, the author will discuss the most important examples of the relationship occurring between the real and virtual space and a forecast of their cultural and functional consequences. The theoretical basis of the observation and research conducted by the author is the ideas referring to the space of flows by Manuel Castells contained in his work "The Rise of the Network Society" (1999) and the views expressed by John Thackara in "In the Bubble; Designing in the Complex World" (2006). Theoretical works systematize knowledge on the impact of information on society and confirm the observations that the universal access to it is one of the most characteristic features of our times, and it is a fundamental element of culture formation. Under this influence, architecture is changing in the sphere of concepts, in the field of design and at the construction stage. The perception and communicative functions of architecture have been significantly transformed.

The theoretical basis of the work is construed upon publications of sociologists and economists who have been depicting a "society of the future" for several decades. The below publications describe the changes of society, and their authors treat the occurring spatial changes in terms of the secondary consequence of the described state of affairs. Their considerations are, however, the basis for the claim that the appropriate shaping of space can help in the creation of the information society. The authors of the particularly inspiring works include:

Toffler (1980) who called the projected transformation as the "third wave"—in comparison with the first wave, which was a civilizational transition to the agricultural system, and the "second wave", brought by the industrial revolution. The picture of a society of the future outlined by Toffler in 1980 has not become outdated, and we can still treat it as one of the most important opinions on forming a society of the future. The author depicts possible extensive cultural, social and economic changes: changes in the family, pluralism, employment, education system, the end of national states, new forms of corporations. Graham (1995) perceived and described the similarities between a real space and a digitalized world, whereas Jeremy Rifkin defines the coming era as the "age of access" claiming that the most important value is the possibility to use, access to information and goods.

Several authors, however, include information on the expected impact on the surrounding space: Elżbieta Wysocka in her published work "The Impact of the Information Society on the Spatial Planning" took up the studies on the impact of the development of information technology on the development of geographic regions. The theme of the impact of the development of information technology on a city scale was undertaken by subsequent authors.

In 1995, William Mitchell published a now classic book "The City of Bits", in which he compared the actual urban spaces and places in cyberspace. According to his concept, virtual and real spaces overlap each other—and they will be able to interweave, replace and support one another. According to Mitchell, the impact of information technology will be an important factor in the urban design. This new form is called "E-topia" (Mitchell 1996).

In Poland, this topic was addressed by Anna Palej in the book "Cities of Information Civilization. Looking for a balance between the physical world and the virtual world" (2003) and by Jan Słyk in "Sources of Information Architecture" (2012).

Castells (1999) dedicated one chapter of "Network Society" to influence of the "space of flows", that is, among other things, the development of information technology on the phenomena associated with the evolution of architecture. In the chapter "The Architecture of the End of History", he defines spatial phenomena in the following way: "manifestation of the dominant interests takes place around the world, and across cultures, the uprooting of experience, history and specific culture as the background of meaning is leading to the generalization of ahistorical, acultural architecture". Castells only points out that topic—in relation to the realization by Steven Holl. Nonetheless, he does not develop it into a broader discourse.

Rem Koolhas talks about adapting architecture to delocation. He sees Lille Grand Palais Convention Centre as the expression of the "space of flows".

In contrast, John Thackara in "In the Bubble: Designing for a Complex World" pays attention to both the architecture and the creation of innovative solutions in designing utilitarian objects (*design*).

In summary, the above-mentioned authors agree on the principles of changes in information and communication technologies that affect the spatial environment. They underline the changed role of education, the model of space and work in a unified space.

References

Biggs, M., & Buchler, D. (2008). Eight criteria for practice-based research in the creative and cultural industries. *ADCHE, 7*(1), 5–18.

Bonenberg, A. (2013). *Media, przestrzeń, architektura. Transformacje przestrzeni społeczeństwa informacyjnego*. Poznań, Wydawnictwo Wydział Architektury Politechniki Poznańskiej.

Briggs, A., & Burke, P. (2010). *Społeczna historia mediów*. Tłumaczenie z angielskiego Jakub Jedliński. Warszawa: Wydawnictwo Naukowe PWN.

Castells, M. (1999). *The rise of the network society*. Oxford MA: Blackwell.

Celiński, P. (2010). *Interfejsy: Cyfrowe technologie w komunikowaniu*. Wrocław: Wydawnictwo Uniwersytetu Wrocławskiego.

Chomsky, N. (2002). *Media control the spectacular achievements of propaganda*. NY: Seven Stories Press.

Graham, S. (1995). Cities, nations and communications in the global era: urban telecommunications policies in France and Britain. *European Planning Studies, 3*(3), 357–380.

Grant E. (1996). Średniowieczne podstawy nauki nowożytnej, Warszawa 1996. [Original title: The Foundations of Modern Science in the Middle Ages] Grant E., Średniowieczne podstawy nauki nowożytnej, [Original title: The Foundations of Modern Science in the Middle Ages] Translated by T. Szafrański, Prószyński i S-ka. Warsaw.

Horkheimer, M., & Adorno, T. W. (1979). *Dialectic of enlightenment*. London: Verso.

Lévi-Strauss, C. (1993). Spojrzenie z oddali, Warsaw.

Lichański, J. Z. (2000). Retoryka. *Od renesansu do współczesności—tradycja i innowacja*. DiG. Warsaw.

Luhmann, N. (2007). *Systemy społeczne: Zarys ogólnej teorii*. Tłumaczenie z niemieckiego M. Kaczmarczyk. Kraków: Zakład wydawniczy Nomos.

Manovich, L. (2011). *Język nowych mediów*. Tłumaczenie z angielskiego P. Cypriański. Warszawa: Oficyna Wydawnicza Łośgraf.

May, R. (1997). *Błaganie o mit*, [Original title: The cry for myth] Zysk i s-ka, Poznań.

Mitchell, W. (1996). *City of bits: Space, place and the infobahn* (p. 48). Cambridge: MIT Press.

Musioł, S. (2012). *Symbol, mit, rytuał w komunikacji marketingowej*, Lingua ac Communitas. (Vol. 22, pp. 169–180) (ISSN 1230-3143)

Nierenberg, B. (2001). *Zarządzanie mediami: Ujęcie systemowe*. Kraków: Wydawnictwo Uniwersytetu Jagiellońskiego.

Skąpska, G. (2007). Niklas Luhmann i teoria systemów społecznych, w: *Systemy społeczne. Zarys ogólnej teorii*. Tłumaczenie z niemieckiego M.Kaczmarczyk. Kraków: Zakład wydawniczy Nomos.

Thackara, J. (2006). *In the bubble: designing in a complex world*. Cambridge: MIT Press.

Toffler, A. (1980). *The Third Wave*. Toronto: New York: London: Bantam Books.

Chapter 2
Architectural Concept: Location Versus Tradition of a Design Philosophy

Abstract In globalized world, the geographical location of the buildings is loosing its impact on architecture, as places are loosing their identity. What instead begins to play an important role is an origin of the design concept behind architecture. It is largely determined by the creative individuality of a designer, which, in turn, is never a completely independent—self-induced feature. What should be pointed out here is the crucial role of architectural education. In the globalized architecture, academic tradition is behind the actions of the creator: design philosophy, understanding and interpreting the design process, sensitivity, ethics.

Defining the territorial framework in the research is commonly justified by the belief that the location of the building is the primary factor influencing the characteristics of its architecture. This issue is closely connected with the theme of the spatial context—understood as the direct environment of the building and the cultural context of its creation. Contemporary architecture is subject to currents of the global flow of information, becoming part of Castells' "space of flows" (Castells 2007a, b). Its formal features are decreasingly dependant on tradition, spatial and geographic context, and they are increasingly imprinted with global media patterns and design trends. In the description of architecture, which is global in its nature, numerous examples of which are cited by the author of this work, in addition to the geographical coverage of the developed content, one should also pay attention to:

1. mobility of architects,
2. internationalization of the creative environment,
3. internationalization of the architectural education.

Adaptation of architecture to the surrounding conditions over the centuries has had various causes and consequences. Traditionally, they included availability of technology, materials, climatic conditions, cultural context. Modernity has brought mastering of technologies allowing to build similar forms in various conditions, unification and globalization of the market of materials, making the builders independent from resources of the specific location. Since the nineteenth century the discourse on architecture and spatial context has been expanded by the doctrines

A. Bonenberg, *Cityscape in the Era of Information and Communication Technologies*, The Urban Book Series,
https://doi.org/10.1007/978-3-319-69542-6_2

of conservation issues, which implied a conscious adjustment, contextualization of newly designed objects. Not all theorists and practitioners continue this approach. At the same time, in the centre of this discussion is the creator—the architect and his personal relationship with the place. The question is what the project is better defined by—the tradition of architectural thought, philosophy from which the creator is derived or the place—*locus*—is the most appropriate.

Determination of the geographical boundaries in the research on architecture is considered to be a fundamental issue. It is often understood as equivalent with giving information about the identity of the place, tradition, culture of creation and implementation of urban or architectural design. This approach is indeed true when dealing with issues of historical architecture. The place defined the availability of materials, materials defined the building technology, which in turn affected the applied architectural detail. Historical architecture was obviously subject to a stylistic evolution and international influences—the locations were, however, so much separated from each other that the influence progressed relatively slowly. Over the years the patterns that came from afar were changing, mixing with the native stylistic solutions. And these objects, inscribed in the landscape and fabric of the city, were studied by the first students of the architecture courses and schools in the past centuries. Drawing on existing models formed the basis of research methodology associated with the beginnings of the architectural education. The study was based on the analysis of objects in an authentic context during European journeys underlying the architectural education. Sketches of construction forms and details made by the then students survived, inter alia, in sketchbooks and documentation preserved till now. A perfect example is a prominent Scottish architect from the Art Nouveau era, Charles Rennie Macintosh (Robertson 2002). A careful study of observation of context and coupling of architecture with the surrounding space in the case of his works arouse admiration—they explain the criterion of a place and cultural context as a basis for discussion about the quality of historical architecture.

Today, however, the increasing number of realizations elude to be classified by location. Historically, the creator of the work was usually associated by his or her origin with the community which decided to erect an architectural object. In the case of architects—visitors from foreign countries, the country of conducting investment was at least a temporary place of residence of the creator—so he was no stranger to the social environment. Culture, language, traditional places influenced the authors, encouraging them to borrowing and creative use of local motifs. The need for the use of local building materials, the use of the services of local craftsmen in the construction process—led to the mixing of technologies for erecting of buildings. Spatial concepts and decorative motifs could be imported—for example, a wonderful renaissance of Cracow—despite the Italian roots—it had also native features. Nineteenth-century architects working in Cracow, who finished schools in Vienna and Munich, also imported styles processing them, however, in their own way—and under the influence of the local culture. For these reasons, defining a territorial framework for objects of great historical architecture provided information on a wide cultural context of the work.

Today, the range of activities of creators has been significantly extended. Restrictions for beginning works on the project no longer lie in the geographical distances, they were substantially limited by the widespread use of information and communication technologies (ICT), and the availability and generalization of air transport. Locations of new design projects are of secondary importance, also due to the tradition of contracting local architects[1] to carry out some design work and supervision. The flow of duties and commands does not need a hierarchical order—it is more like a collaboration of entities involved in a common process of realization. This solution is more natural since architectural offices have always had the problem of coordination of many industries—the institution of local architects is another link.

The possibility of quick displacement limits the impact of culture of the location on the Creator, and, unfortunately, it also reduces his or her knowledge of that place. An architect evaluates the characteristics of the surrounding area from the point of view of the visitor—tourist. Interpretations of the local architecture, which are often the element around which the marketing framework of a new investment is built, may prove to be shallow and borrowings—superficial. This is the case, in the opinion of the author, of the international airport in Beijing, designed by Norman Foster. The motif of the red columns visible in pictures, dominant in the external composition of the solid and in its interior along with a symmetrical system are present, among others, in the architecture of the Forbidden City. Formal quotes, the use of symbolism of colours, are very direct and obvious. They are based on a simple processing of the formal characteristics of the most popular monuments in China.

Similar situations occur frequently and result from expectations of *consumers* of architecture. The latter are interested in the simultaneous stressing the identity and culture of the place and in belonging to an elite global network of business, economic connections, whose style is synonymous with modernity and progress. These objectives are not always moving in one direction. They embody the contrast occurring between what is global and local. An investor with public capital is usually eager to highlight the references to the past and tradition because it is required by the culture and the local community, big business is looking for references to global design and universal themes. Nonetheless, both the authorities of bigger and smaller towns think that having a building bearing the name of a well-known creator of world architecture opens the door to an exclusive club of big cities—it is an indicator of belonging to the "great world"—which stands in opposition to the sense of familiarity and identity of the place (Figs. 2.1 and 2.2).

An important element in determining the nature of the global market architecture is a multiculturalism of creative environments. As in other areas, the custom of employing a multi-national project team serves to prepare for operations in foreign markets. The basic requirement is knowledge of the specificity associated with conducting business in other countries and cultures. Employment of people speaking the language of the place for the project realization, people with knowledge of the legal regulations, cultural specificity is increasingly becoming a necessity—like

[1]*Local architects*—Architectural office located in the project realization site.

Fig. 2.1 Buildings designed by the architectural office Steven Holl Architects: linked Hybrid apartment (Beijing, China, 2009), Simmons Hall, Massachusetts Institute of Technology (Cambridge, USA, 1999–2002). In these projects strikes the similarity of spatial solutions, formal solutions and architectural details. *Photo* A. Bonenberg

employment of specialists to perform specific tasks. Impact of the creative environment on the result of the design work is significant. In the international team of designers occurs the exchange of experiences, both professional and philosophical ones. Both elements affect the applied spatial solutions. Shaping of meeting places, interaction space, working space, residence depends principally on traditions, culture and beliefs. In the project teams we deal with merging and imposition of frequently very different ideas. The privacy issue or family model are variously perceived in diverse places—and these factors influence the development of residential space. The business culture is also different—hence the different schemes of working space. The author experienced these differences working in the office Renzo Piano in Genoa, in an international creative environment (Figs. 2.3 and 2.4). Apart from having to choose concepts that were satisfactory to a designer, there was the dispute of the geographical diversity of implementation and issues relating to local climate and local legislation. The objective of establishment of international project teams in was:

Fig. 2.2 Simmons Hall, Massachusetts Institute of Technology, Cambridge, USA. *Photo* A. Bonenberg

1. obtaining the adequate statement of skills: professional, language competence.
2. assignment of roles in which the specific qualifications are useful,
3. ensuring and facilitating the possibility of proper communication in the team—including the linguistic dimension; employees should be able to speak a common language and should be encouraged to a permanent use of it on the group forum. Lack of such solution reduces the efficiency of the group.

A large-scale professional mobility of architects is confirmed by statistical data contained in the atlas of architecture by Phaidon.[2] Data on global relationships between architects and implementations indicate a particularly high proportion of architectural offices located in Europe in shaping the face of a world architecture. Out of 653 cited authors, 323 operate on the Old Continent, at the same time, out of

[2]"The Phaidon Atlas of twenty first Century World Architecture" page 10. The statistics in this publication can be considered reliable because of the comprehensive scope of the subject elaboration. Accurate, because the book is the quintessence of world architecture.

Table 2.1 Expansiveness of architectural offices

Europe 323 design offices	• 154 intercontinental projects
Asia (a large dominance of Japan) 113 design offices	• 33 intercontinental projects
North America 97 design offices	• 37 intercontinental projects
Oceania 43 design offices	• 4 intercontinental projects
South America 52 design offices	• 3 intercontinental projects
Africa 25 design offices	• 3 intercontinental projects

Source "The Phaidon Atlas of Twenty First Century World Architecture", page 10

233 researched intercontinental projects, as many as 154 were implemented by the European offices (Table 2.1).

Table 2.1 shows that Europe is unrivalled in terms of influence through export of design concepts, the subsequent places are occupied by Japan and North America.

Fig. 2.3 The architectural office Renzo Piano Building Workshop: facade, teamwork space on the lowest terrace of the office. The glass facade and sliding wood panels used for exhibition of prints and drawings which enables discussion of the project in a broad group of people. *Photo* A. Bonenberg

Fig. 2.4 Renzo Piano Building Workshop, model workshop. The construction of mock-ups is an essential design tool of Renzo Piano. *Photo* A. Bonenberg

In the situation when the location of the project has been losing its importance, what begins to play a more important role is an origin of the design concept. It is largely determined by the creative individuality of a designer. This, in turn, is never a completely independent—self-induced feature. What should be pointed out here is the crucial role of architectural education. In the global architecture, tradition has the greater importance than a geographic location, since tradition is behind the actions of the creator: understanding the design process, sensitivity, ethics, sensitivity to context. What has a significant impact on the quality of the international architecture in the era of the dominance of Western culture are precisely universities—schools of architecture. In the study conducted among the students of the Faculty of Architecture of the Technical University of Poznan 98%[3] of respondents indicated a strong influence of the "master"—teacher on the effects of their own, individual creativity. According to the author tracing and examination of the "creative thought"

[3]A statistical survey was conducted on a sample of 430 students in two age groups—in the second year and fourth year of study during the summer semester 2011.

gives us a better idea of the circumstances and philosophy behind the creation of a work of architecture than the use of a geographical key. University centres, through the selection of staff, international cast of professors, elaboration of the adequate, up-to-date and modern *curriculum* of the taught subjects and implementation of the educational process, bear part of the responsibility for the quality of the designed space and globalization of the stylistic trends (Bonenberg 2013). Their basic duty is to shape a responsible attitude in relation to:

1. past: heritage of the spatial context,
2. present: concept, philosophy, available technologies, adequacy of material solutions,
3. future: adoption of issues of variability in architecture.

An important feature of European and American architectural education centres are mutual contacts and cooperation. It can be said that the process of architecture globalization begins at the stage of education of a student. A student who participates in the universal international exchange programs—for example (Erasmus[4]). A student taught by professors cooperating with other centres—conducting joint scientific works and research. These trends affect the unification of a scholastic culture—a globalization which is later visible in the architectural realizations. Approximation of teaching methods, level of qualifications is a positive phenomenon. It greatly supports and facilitates the international flow of ideas and skills.

A degree of internationalization of education centres is one of the important criteria for assessing the quality of teaching.[5] Centres offering a high level of education receive international students. Their presence and further professional path widens the impact of the academic centre on the further practical and theoretical development of disciplines—it provides a broad impact on an international scale and raises the prestige of the University. The language issue is also one of the reasons why the culture (including architectural culture) of Anglo-Saxon countries is very influential.

One of the oldest and most famous schools of architecture is the Architectural Association in London. Graduates of this school with an unusual program and a flexible approach to knowledge include, among others, Rem Kolhaas, Zaha Hadid, Peter Cook, Peter Wilson and Julia Boles-Wilson. Analyzing achievements of those few outstanding personalities of a contemporary architecture we can realize a wide intellectual range of the academic community—teachers and students.[6] The author had the opportunity to participate in the evaluation of teaching methods of

[4]In the international student exchange of Erasmus Programme (European Community Action Scheme for the Mobility of University Students) has been taking part generations of students since 1994 (Socrates programme), there have been conducted ongoing activities aimed at integration and cooperation within the European Community.

[5]The criterion used as one of the six in the University Academic Ranking of 2011, source: "Rzeczpospolita" (the Polish newspaper).

[6]Both Rem Kolhaas and Zaha Hadid, after obtaining diplomas, were academic teachers at the Architectural Association School of Architecture in London.

Fig. 2.5 Globalization in the architectural education: towards new designing forms and technologies. The inner courtyard of the London Architectural Association School of Architecture with a parametrically designed cloud installation spread out over the terrace

Architectural Association, where a very individual approach to students is applied. Systematics and distribution of courses are characterized by arrangement in blocks compared with the most European centres.[7] The unusual curriculum promotes the formation of avant-garde projects—with use of advanced computational methods being the primary design tool (Figs. 2.5 and 2.6).

[7]A coherent system for dividing courses was developed in Europe on the basis of ECTS points, Architectural Association, School of Architecture in London does not belong to that system.

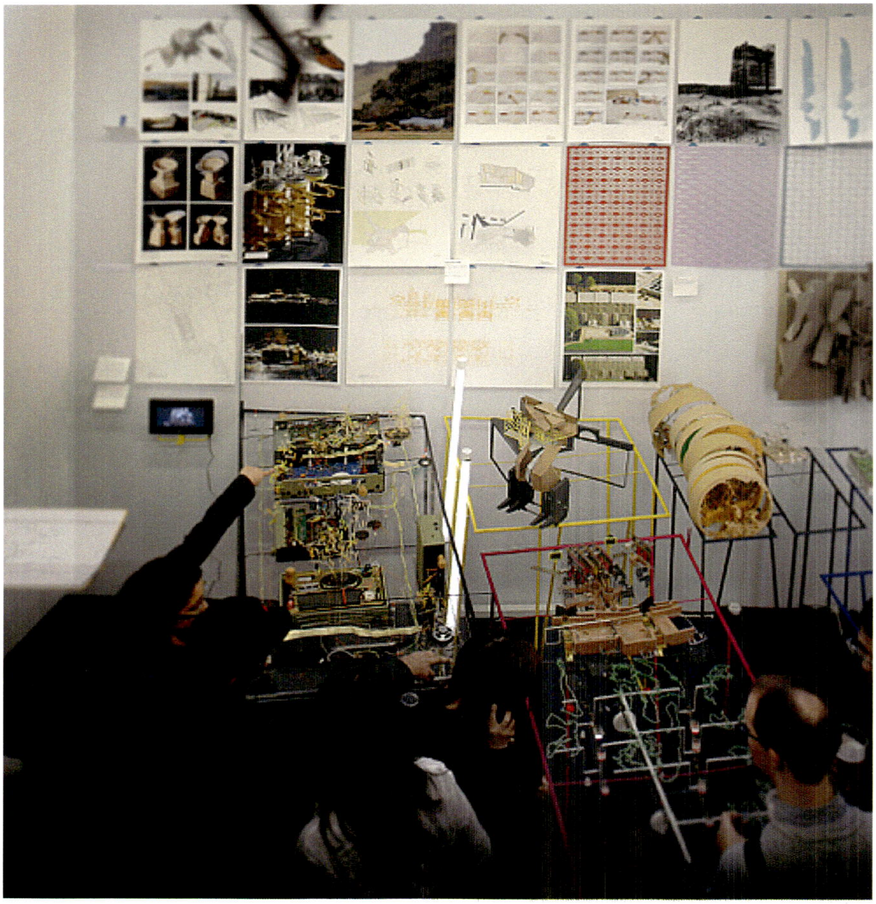

Fig. 2.6 The annual exhibition (*summer show*) of the recognized London School of Architecture Bartlett. *Photo* A. Bonenberg

The statistical data relating to the architectural education in Europe show the disparities in the amount of students of architecture being formed and the number of practicing creators. It seems that this condition causes an asymmetry in the availability of architectural services and encourages to a professional migration. These disparities are significant in Europe. At the moment, the state with the highest number of students of architecture in Europe is Italy. A characteristic feature of the Italian architectural education is the emphasis on design theory and its *ex cathedra* relationship to the history.[8] It seems, however, that the north-European approach

[8]Diagnosis based on the long-term cooperation with the Dipartimento di author Progettazione dell Architettura, Politecnico di Milano.

Table 2.2 Disparities in the amount of students of architecture being formed and the number of practicing architects in Europe

ITALY	97,200 students of architecture	194 practicing architects/100,000 inhabitants
SPAIN	21,200 students of architecture	93.9 practicing architects/100,000 inhabitants
PORTUGAL	9300 students of architecture	120 practicing architects/100,000 inhabitants
(…)	(…)	(…)
POLAND	6,700 students of architecture	35.3 practicing architects/100,000 inhabitants
(…)	(…)	(…)
LICHTENSTEIN	120 students of architecture	261.6 practicing architects/100,000 inhabitants
ESTONIA	105 students of architecture	54.1 practicing architects/100,000 inhabitants
ICELAND	36 students of architecture	112.5 practicing architects/100,000 inhabitants

Source "The Phaidon Atlas of twenty first Century World Architecture" page 227

brings better results in the architectural education, and it can be described as *hands on experience*, or a practical verification of an obtained knowledge (Table 2.2).

The role of universities is important in the information society. Hirst (1995) in the article "Education and the Production of New Ideas" draws attention to the phenomenon of loss by universities of a monopoly on the transmission of information. There have been created many competing branches of economic activity devoted to this task: training courses, specialized consulting, education in the form of multimedia communication. Schools, which have traditionally played the role of information transmission—are facing the necessity of a partial redefinition of their role in the information society. Their opportunity is the fact that mass media do not create knowledge—they only recycle it and provide information Universities, including architectural ones, should take the role of a producer of knowledge. Critical thinking, objectivity of research, innovation—herein lie the possibilities of academic centres. What has turned out important is the development of theoretical research, prototyping—creating knowledge and verifying the results. At the same time, their role as a transmitter of encyclopaedic information and images is decreasing. Images of the global architecture we can see in many sources: with the eyes of thousands of users of architectural blogs, resources of Picasa Albums (great pictures), finally, from a bird's eye view in Google Earth. Such information can be the basis of the space analysis, but the knowledge drawn from them is superficial. A student basing only on that knowledge has a tendency to create aesthetic clichés, duplicate formal and material solutions. The architectural education should be focused on creating innovative, insightful solutions. Designs, in some of its part, should look to the future, teaching to young architects options of adaptation, response to the real needs of users.

Fig. 2.7 DRL 10 Pavilion, Bedford Square, Architectural Association School of Architecture, London. An example of a digitally generated small architecture: a spectacular shape and experimental building material. One cannot receive it only as an exercise—a form of the object is very conducive to social behaviours, meetings during lunch breaks. The pavilion is a landmark in the cityscape. Drawing A. Bonenberg

So, if it is not the matter of connection with the place that can provide information about the philosophy, culture of creation, identity and tradition behind the work belonging to the "global architecture"—what, therefore, can?

Under the conditions of the above-mentioned high mobility of students, creators of architecture and activity of architectural offices, we know much more about the architectural work observing the tradition of a creative philosophy than analyzing its geographical placement, its location. A contemporary architecture has become less dependent on the place of origin both in terms of available technologies as well as materials and concepts. Globalization in architecture is a consequence not only of interpenetrating global stylistic trends, but also of mobility and multiculturalness of the environment of creators. The environment whose roots date back to the European and North American traditions of architectural education (Fig. 2.7).

References

Bonenberg, A. (2013). *Media, przestrzeń, architektura. Transformacje przestrzeni społeczeństwa informacyjnego*. Poznań, Wydawnictwo Wydział Architektury Politechniki Poznańskiej.

Castells, M., 2007. *Społeczeństwo sieci*, [Original title: The Rise of the Network Society], 412. Warsaw: Wydawnictwo Naukowe PWN.

Castells, M., 2007. *Społeczeństwo sieci*, [Original title: The Rise of the Network Society] Wydawnictwo Naukowe PWN, Warsaw. 412 a. **412**.

Robertson, P. (2002). *Charles Rennie Mackintosh Architectural Sketches*. Glasgow: Campsie Litho Ltd.

Hirst, P., 1995. *Education and the production of New Ideas*. AA Files 29, London.

Chapter 3
Mass Communication and Public Space

Abstract Traditionally, urban public spaces integrated inhabitants of a city. It provided the setting for the "space of access to information", and this characteristic was often expressed in the architecture with elements of visual information: inscriptions, sculptural detail of the "communicative" role. Symbolism of hierarchy was expressed by the dominants of sacral and secular buildings. As mass communication takes over, the fact of "detachment" of communication from a real space becomes a fact, and it remains not without consequences on the space.

Standing in the centre of the Roman Capitol, at the end of the Via Sacra, a visitor to the city will be in front of the ancient Rostry platform. A place of great importance, the heart of the Roman Republic, the heart of the empire, from which new decisions of governments and news spread to the entire ancient world, which occupied the most important place in the composition of the Forum Romanum. In fact, the seriousness of the place and function has been exposed, highlighted with prows of captured ships—both Via Sacra and Rostra grown to a symbol—the heart of the ancient empire.

The spaces of streets and squares in cities since the beginning of recorded history of settlement were delineated in order to create a public space—the space of social integration (Figs. 3.1 and 3.2). The primary role, task and cause of the existence of the common areas in the city are the need to create an urban community—in order to improve the political organization of inhabitants (Kostof 1992a, b). The binder of the community is a shared access to information, a possibility of easy communication of it and taking action in response to it. Building and maintaining relationships between the residents is the primary factor that allows for the spatial, economic, political and cultural development of the city centre. Without the integration of a community, there would not be a chance of working as a collective, and the arise of complex social relations so characteristic for modernity would be impossible. To this day, the projects of revitalization of cities, districts and centres are primarily associated with the creation or restoration of the local community associated with the place.

In the historical times, access to information was interpersonal, direct and contributed to the preservation of one of the basic human traits—a man as a social

A. Bonenberg, *Cityscape in the Era of Information and Communication Technologies*, The Urban Book Series,
https://doi.org/10.1007/978-3-319-69542-6_3

Fig. 3.1 Public spaces of medieval cities: a living and trading place do not meet the representative functions. Narrow streets (viccoli) in the town of Albenga (Liguria, Italy). *Fig.* A. Bonenberg

being with all the psychological and behavioural consequences. The need to live in the community lies in our nature.[1] Technical inventions such as printing techniques (fourteenth century), telegraph (seventeenth century), telephone (nineteenth century), radio (twentieth century), television (twentieth century), the Internet

[1]Aristoteles as the first wrote about a human: *Man is by nature a social animal; an individual who is unsocial naturally and not accidentally is either beneath our notice or more than human*, "Politics", about 328 before Christ.

Fig. 3.2 Public spaces of Baroque cities. Corso Vittorio Emanuele, Noto, Sicily. *Photo* A. Bonenberg

(twentieth century) made that access to both the global and the local information stopped being based on interpersonal relations. Social contacts have changed their nature and enriched with new tools. Over the centuries, access to information became more and more loosely associated with a direct contact between people. In this context, the next step, which is building of the information society, is in some sense a critical moment in history: it is possible to work, have fun, function in society and build a relationship with another person—with a very significant reduction of direct contacts. Questions about the role of direct contacts in social organization are fully justified. To what extent will follow their self-limitation? To what extent will we be ready to replace a face-to-face conversation with other forms of communication? Will the participation in social and mass communication satisfy human social needs?

When considering the transformation of space for the information society, we shall ask what is the future of a city? What will happen to its public spaces? Will there be a need to include them in new projects?

The temptation of partial closure in virtual contacts exists, and there are also opportunities for work, entertainment only by the network. At the same time, we should pay attention to the value of direct interpersonal contacts, as a factor of building relationships and the therapeutic element. During the conversation, the non-verbal communication, after all, exceeds the effect of linguistic communication, and it is most difficult to convey it through digital means of communication (even with image transmission). Most likely, therefore, the personal contact with others is necessary to achieve life satisfaction and happiness. Referring this observation to designing of the space, there is a need and a mission of building the space for social integration—for the information society.

In "Life Between Buildings" Gehl (2009a, b) writes about the three types of population activities in public spaces: necessary activities, optional activities and social behaviours (Gehl 2009a, b). The necessary activities are the ones which must be performed in the public space. The optional activities are the ones that a user of the space performs only under certain favourable conditions, most often in areas with high values of composition and landscape. *Social behaviours are those that depend on the presence of other people in public spaces. (…) Social behaviours occur spontaneously as a direct result of movement and staying of people in the same space. This is due to the fact that social behaviours are favoured (though not directly) by the situation when necessary and optional activities have better conditions in public spaces* (Gehl 2009a, b).

In the information society, access to information in the form of direct communication changed its classification, in the classification cited above. From the category of necessary actions made in the city space, in historical times, it turned into almost exclusively social actions (and therefore optional)—in the present.

In the historical period, obtaining information was for the majority of the population inseparably associated with entering the urban space. It was necessary for undertaking economic activities and participation in a political life. The effectiveness of community integration was associated with the composition and development of public spaces. In the ancient, medieval, renaissance, baroque and classicist cities, the basic components of the space were similar: clearly arranged interior sequences were created, and their system was characterized by a high degree of customization and clarity. The individual elements of the structure were of diverse hierarchy and destiny—from representational to strictly utilitarian ones. This order contributed to the integration and organization of various groups and social classes, optimizing the transfer of information within them, facilitating coordinated actions and an internal organization (Figs. 3.3 and 3.4).

Traditionally, the public space integrated members of religious groups, ethnic minorities, and members of the craft guilds. The setting for the "space of access to information" was and is the architecture, enriched with signs of visual information: inscriptions, sculptural detail of the "communicative" role. Public space represented and still represents the setting for daily activities, hence the message of the symbolism expressed by the dominants of sacral and secular objects, in the ornamentation of buildings belonging to the frontage of the streets and squares, is important and it has been present in the lives of inhabitants of cities till today. Currently, some of the historical meanings have become unreadable due to the ongoing cultural changes and migration—breaking of the link between a community and a space. To this day, in the public space, there are its unique medieval and renaissance elements

Fig. 3.3 Elagant square Piazza Duomo in Catania (Sicily) is used during major state and church holidays; in the picture is shown a boisterous celebration of the city patron day, Saint Agatha. The social integration refers to the entire urban community, visitors, tourists; it is associated with the politics and worship. Buildings—great monuments of Baroque architecture along with an impressive elephant fountain at the central point. *Photo* A. Bonenberg

for announcing messages delivered by messengers and merchants. They are recurrent in diverse forms in the markets of European cities: from the far north and Scottish *mercat* cross in the form of monuments to the decorative balcony Broletto in Como.

Until the nineteenth century, the ways of communication and the space of streets, highways, transport pathways overlapped and coincided. It is difficult to determine what lays at the basis of the original motivation for finding new ways—was it the desire of profit through trade contacts? Or maybe the first was curiosity—and, therefore, access to information? The first inventions allowing for remote transmission of information—radio, telegraph, telephone—separated the function of access to information from the physical space, but their extent was so limited that their impact on the use of space was minimal. A number, "weight" of information transmitted in this way, was limited. Only a very considerable increase in data transmission capabilities, low cost of use and broad popularization of information and communication technology changed the situation. The existence of a non-physical space of the exchange of messages and communication, as well as

Fig. 3.4 A usable space of the fish market in Catania (Sicily). The fish market in the area of Via Pardo is open every day. In the heat of the Sicilian climate, the smell becomes a considerable obstacle for this function. The social integration occurs between persons directly involved in the trade: the tourists come here occasionally. The surrounding historical buildings are modest and neglected, and the amount of the elaborated architectural detail—much less than in the Piazza Duomo. *Photo* A. Bonenberg

social and cultural changes that are its consequence, and that are still occurring, affects the important design decisions concerning the development of the composition of cities, public space and architecture. Traditional functions, hierarchy and character of the space are changing.

In the nineteenth and twentieth centuries, the urban development of cities was associated with improvements and optimization of transport systems. In 1894, Arturo Soria y Mata proposed a plan of a Linear City, "Ciudad Lineal" based on a wheeled transport, pedestrian routes and primarily, the construction of public transport lines, including a proposal for making the underground transport.[2]

The design of an industrial city by Toni Garnier, Cité Industrielle, presented ten years after the Linear City, postulated a functional zoning of area for: work, social life and living zones. Such a system enforced a solution of the hierarchical communication system—in order to connect these zones. In the middle of the system, there was outlined the commercial-service-administration centre constituting the main space for a social integration. Low buildings, large areas of green and pedestrian communication routes were a reference of the project to the concept of the Garden City by E. Howard (Wujek 1986; Nebois 2003a, b).

A significant departure from the historical rules of shaping urban space occurred along with realizations of projects of cities in the age of modernism. In addition to the significant spatial growth of urban centres, there was adopted the idea of a functional division into industrial, residential, recreational areas, what contributed to the development and importance of the communication infrastructure. Transformations contributed to increasing mobility of residents within the city—the spread of information in the direct form was faster and more efficient thanks to transport, and it was enhanced by the participation of the mass media. Significant effects accompanied the separation of wheel and pedestrian communication in accordance with the principles of the Charter of Athens. Experimental projects, which were the basis of these ideas, included the "Voisin" plan for Paris (1925) by Le Corbusier, showing 24 skyscrapers organized around the node—the communication centre. This concept is characterized by functional separation and a significant share of greenery. "Obus" (1931)—the project of development of Algiers—presented the overwhelming predominance of infrastructure associated with the automotive industry (Nebois 2003a, b).

More less at the same time, Frank Lloyd Wright, "Broadacre City" (1929–1935), was working on a different topic concerning a city, namely concepts of development of the emerging American suburbia. The solution was closely linked to sociopolitical ideas postulated by Wright. Residents were to obtain from the Federal Reserve 1 acre of land, mainly for agricultural activities, and the local community was supposed to be acting on the principles of a cooperative. The system had a repetitive character, fully based on a road transport (Levine 1996).

In these projects, the most important is the question about the nature of space for direct transfer of information, identified with the spaces of social integration. They were deprived of the classical hierarchy encountered in historic cities. Significant here is the large scale of assumptions and the lack of an individual character due to the repeatability of the composition. These problems became visible with the implementation of settlements under the new assumptions, with the development and popularization of the idea of mass living. The attractiveness of the concept of

[2]It was a realistic proposition because the first underground railway in the world, the London underground, had already been used then for about thirty years. The first section of the subway was opened in 1863. Therefore, author's ideas went hand in hand with the technical capabilities of the era. The premises of the Linear City's authors can thus be classified as real.

mega structures as the basic components of the city, in theory, provided an opportunity for infinite spatial expansion and unrestricted development.

One of the precursors of the urban design based on repetitive elements of a mega structure was Kenzo Tange. He proposed in his plan for Tokyo a large-scale solution based on residential and commercial units. The plan was aimed at structural reorganization and expansion of the infrastructure of Tokyo in the direction of the Tokyo Bay in the form of "civic axis" (Brayer et al. 2005a, b, c, d, e, f, g). In that time, Europe predominated the projects aimed at creating a new "urban landscape": "Motopia" (Jellicoe 1961) and "Crater City" by Yona Friedman developed in 1961. Studies on the evolution of landscape, proposals of a city in which cars would be moving on flyovers, suspended roads and helicopters would be the main means of transport were suggested by Jellicoe. And Yona Friedman built an "artificial landscape" which would enable each resident to change the location of his or her own home. The project was a combination of the idea of "social mobility" and mega structure (Brayer et al. 2005a, b, c, d, e, f, g).

The ideas of an endless city were analysed in the works of Eilfried Huth and Günter Domenig. A particularly significant project was the concept for Ragnitz in Austria: the idea of an infinite expansion, modularity, openness and flexibility of the three-dimensional composition meant a lot of planning freedom. The project also became a starting point for architects to create structures with varying geometry, mobile ones (Artefix Project, 1970–1972). The momentous epoch-making technological discoveries included in the table encouraged to create a futuristic visions of urban space. Thanks to them the building ceased to be a single architectural object and became a system of environments arranged in an infinite composition. Realizations were characterized by a low degree of customization and clarity. The traditional code of information transfer on location contained in the traditional composition of the city ceased to apply. Lack of a defined integration space discouraged a direct exchange of information. The possibility to maintain a complete anonymity limited the social integration and building of a neighbourhood community. The problem lays in the lack of acceptance and agreement to take into account the individuality of an individual and social, neighbourhood groups. The behavioural psychology research shows that users of public space have a basic desire that the place of meetings was "safe" or, in other words, "defined". It should also have individual characteristics allowing for the identification of users with the space. And in the concepts of infinite cities were missing tools to identify and personalize the neighbourhood space. Another aspect is the scale of the discussed assumptions, the scale which due to the physical size exceeded the capabilities of a human perception: the ability of observation and recognition. Lack of these possibilities has an impact on key aspects, such as the sense of security, the sense of understanding, devotion to the elements of space. A psychological aspect of human nature was shifted to the background.

The crisis of the common, integrating areas that would unite communities also occurred in the cities of the New World (USA and Australia). The cause was not, however, a repetitive system of the planned collective residence. Here the reason was the opposite, a small interference in the planning limited to the delineation of tracts for wheel communication and distribution of plots. A dynamic, poorly controlled urbanization led to the phenomenon of an urban sprawl. The city, by paying extraordinary attention to car and mass communication infrastructure, telecommunications infrastructure, hindered to the residents a direct access to information —the one which determines the formation of the community.

Contrary to the above-mentioned implementing reality of the 60s of the twentieth century, it must be emphasized that this was the period of the creation of theoretical models based on social relations. In some avant garde case study ideas shaping urban space was based on the hypothetical relationships between inhabitants. The first widely popularized project was a Spatial City (1958), a concept based on the dynamic shaping of the city, depending on the existence of a bond (verbal information flow) between the inhabitants. The author of "manifesto" was Yona Friedman. Studying the possibilities of building a "space city"—he constructed theoretical models hovering over the existing buildings. The guiding concept of the shape of the city depended on the residents and was based on the existence of a bond and communication between them. Residents had the opportunity of a free transfer of their place of residence (part of the city) to any other location—that is, create or choose the nearest neighbourhood. This concept turns out to be very up to date nowadays and close to a contemporary philosophy of plug and play (Brayer et al. 2005a, b, c, d, e, f, g).

At the end of the sixties, Peter Cook offered a vision of an Instant City, the nomadic, wandering city—developing the concept of a "travelling metropolis". In this concept, architecture gives way to an audiovisual image, mass events and simulating of spatial environment. An important part of the philosophy was the leading role of communication in the city space: a new visual environment was created in which words and images were displayed on suspended screens integrated into the moving objects. Instant City therefore relates to the information network serving for education, entertainment, leisure and calls for the transformation of architecture into the event (Brayer et al. 2005a, b, c, d, e, f, g).

Concepts of a city associated with the trends of deconstructivism were represented, among others, by Bernard Tschumi and Daniel Liebeskind. In 1987, Daniel Liebeskind presented the concept of a "City Edge", the project of the Tiergarten district in Berlin. An expressive, deconstructed form was meant to be a philosophical reflection on the history of Berlin, to transfer information on the history of the war period presented in the abstract, poetic way. The spatial composition was thus formed in relation to the abstract values: symbolism and memory. They define the directions, proportions and characteristics of planes, lines and solids (Brayer et al. 2005a, b, c, d, e, f, g). Bernard Tschumi, however, based his compositions on the physical interpretation of the movement: the sequence of events that are happening in the architectural space. So he also invoked non-physical elements. The summary of the theory elaborated by him is contained in the collection of essays: "Architecture and Disjunction" (Tschumi 1996), and the only completed project having an urban scale is the Park de la Villette in Paris. The composition is based on intersecting and overlapping lines and surfaces on grids. In 1975, Rem Koolhaas published "Delirious New York"—a manifesto concerning the structure of New

York that drew attention to a specific understanding of the urban grid of the city. The author considers it to be a collection of independent islands—monuments. Each block in the building structure of a metropolis can become an independent, highly flexible unit—which features and appearance result from the imposition of many styles, ideologies and functions in a small space, from the overload syndrome (Nebois 2003a, b). This approach assumes anticontextuality in the creative approach. The postulate of departure from the imperative of references to the spatial context is clearly visible in the later urban concepts, for example Steel Cloud Project for Los Angeles West Coast Gateway (Asymptote), in which the deconstructed city was suspended over the existing city (Brayer et al. 2005a, b, c, d, e, f, g; http://www.asymptote.net). The final step is to the computational, parametric and generative design in urban planning: return to focusing on building typologies as landscape elements, borrowing forms from biological systems—forms of life (Hensel and Menges 2006).

Table 3.1 summarizes the selected urban designs in combination with technological developments in the field of transport and information and communication technologies. From that statement emerges a picture of the changes taking place in the spaces of social integration.

The dynamics of investing conditioned by access to information can be observed in the US metropolitan areas. The increase of urban centres due to the concentration of business and administration lasted until the 80s of the twentieth century. Concentration of building followed despite the wide use of the communication media: telephone and telegraph. The official information in the form of documents, the need for personal meetings (no teleconferencing technology) meant that the geographical location in the centre of the metropolis still meant being in the centre of events—essential for many types of services—according to traditional relationships. Trends in decentralization and escape outside the centres became a reality only after the popularization of information and communication technologies with adequate bandwidth. Only this tool improved circulation of documents and contact with the administration so that companies, commerce, offices, technology centres could effectively develop away from the heart of the great metropolis. They obtained space, cheaper land, easier transport accessibility thanks to individual means of transport. Close, effective cooperation between people and organizations became possible despite the geographical distance. In this way, the availability of information from a precondition for doing business became just one of many possibilities. Decision taking about where to live and work begins to depend on where we feel comfortable (Fig. 3.5).

The possibility of a direct access to information is an important guideline in the design of public spaces and common neighbourhood spaces in the city. Public and semi-public space is essential for the formation of neighbourhood communities: the flow of information in them is of a local character. This fact translates directly into building relationships with the place, creating ties with the surrounding space, because it relates to the common use, management and maintenance.

Communication associated with the mass media affects on a completely different scale. It plays an important role in a social integration, although it concerns

Table 3.1 Summary of selected urban concepts in the context of technical achievement, having a significant impact on the efficiency of information transmission

Progress in areas: information and communication technology and transport		Ideas and concepts in the organization of urban spaces	
Popularization of the over ground as the public transport. Construction of the first underground rail: London underground (the first section already in 1863)	Nineteenth century	1894	Arturo Soria y Mata, a linear city plan, "Ciudad Lineal" based on the theme of transport and communications: the concept of widening streets and building public transport lines, the proposal of conducting underground transport (Malisz 1966)
The Wright brothers and the first controlled flight	1903	1904	Toni Garnier, "Cité Industrielle", a project of the industrial city where the most important characteristic was a functional zoning into areas of: work, social life and living. Low buildings, large areas of green and pedestrian communication routes were a reference of the project to the concept of the *Garden City* by E. Howard (Nebois 2003a, b)
Aviation and actions during the First World War and the golden age of aviation (growth of passenger flights) in the interwar period	1914		
Flying over the ocean of Charles Lindbergh	1927	1925	Le Corbusier, the "Voisin" plan for Paris showing 24 skyscrapers organized around the communication node. This concept is characterized by functional separation and a significant share of greenery (Nebois 2003a, b)
Popularization of car transport (a production mounting tape applied by H. Ford in 1913)	Nineteen-thirties	1929–35 1931	Frank Lloyd Wright, "Broadacre City"—the concept of American suburbia was closely linked with sociopolitical ideas. Residents were to obtain from the federal reserve 1 acre of land, and the local community was supposed to be created on that basis. The system had a repetitive character, fully based on a road transport (Levine 1996). Le Corbusier, the "Obus" plan, the project of development of Algiers—predominance of infrastructure associated with the automotive industry (Nebois 2003a, b)

(continued)

Table 3.1 (continued)

Progress in areas: information and communication technology and transport		Ideas and concepts in the organization of urban spaces	
Beginning of the large-scale passenger aviation. Popularization of helicopters which came into use during the Second World War	Nineteen-fifties	1958	Yona Friedman, "Spatial City". The concept based on the dynamic shaping of the city depending on the information flow and the existence of a bond between the inhabitants. The city suspended over the traditional, dynamically changing (Brayer et al. 2005a, b, c, d, e, f, g)
		1960	Kenzo Tange, plan for Tokyo—structural reorganization of space, typical use of mega structure, plan on the Tokyo Bay (Brayer et al. 2005a, b, c, d, e, f, g)
The era of the space conquest—a triumph of technology: Yuri Gagarin, the first man in space	1961	1961	Geoffrey Jellicoe, "Motopia" (1961): studies on the evolution of cityscape, proposals of a city in which cars would be moving on flyovers, elevated, suspended roads and helicopters would be the main means of transport. (Jellicoe 1961)
Commercialization of access to space: the first private satellite makes a transatlantic transmission of the moving TV image	1962	1963	Yona Friedman, "Crater City"—"an artificial landscape" which enables each resident to change the location of his or her home, combined with the concept of mega structure of "social mobility"—urban structures designed for mobility of users (Brayer et al. 2005a, b, c, d, e, f, g)
The era of the space conquest—a triumph of technology and fascination with advanced technology. The beginning of the Apollo program, first launches of unmanned rockets	1966	1965	Wilfried Huth + Gunter Domain Ragnitz, Austria—city structure planned at several levels supported on steel carrying constructions. The idea is to create an endless city, infinite expansion based on modularity, opening of the three-dimensional composition—the freedom of planning (Brayer et al. 2005a, b, c, d, e, f, g)

(continued)

Table 3.1 (continued)

Progress in areas: information and communication technology and transport		Ideas and concepts in the organization of urban spaces	
Intense popularization of mass media, especially television broadcasting	Nineteen-sixties	1968–1969	Peter Cook: "Instant City", the concept of a nomadic, wandering city where architecture gives way to an audiovisual image, mass events and simulating of spatial environment. The manifesto refers to the timeliness of the information broadcast, its presentation in the form of tabloids and pop culture (Brayer et al. 2005a, b, c, d, e, f, g)
One of the most important technological and medial developments was landing of a man on the moon	1969		
Motorola markets the first mobile phone	1973	1975	Rem Koolhaas' manifesto titled "Delirious New York" refers to a specific understanding of the urban grid of the city. The author considers it to be a collection of independent islands—monuments. Each block in the building structure of a metropolis can become an independent, highly flexible unit—which features and appearance result from the imposition of many styles, ideologies and functions in a small space, from the overload syndrome (Nebois 2003a, b). This approach assumes anticontextuality in the creative approach (Nebois 2003a, b)
The beginning of the Internet for commercial use—it is assumed by the author as the starting point of the information society development	Nineteen-eighties	1987	Daniel Liebeskind City Edge Berlin. An expressive, deconstructed form constituting the reflection on the history of Berlin, transfer of information on its war history presented in the abstract, poetic way. Deconstructionism, symbolism, memory (Brayer et al. 2005a, b, c, d, e, f, g)
Conceptual designs of solar-powered means of transportation. Restrictions on car traffic in metropolitan areas because of environmental reasons		1983–1992	Bernard Tschumi, the concept of planning in the spirit of deconstructionism. Space ordered on geometric grids according to three systems. The focus is placed on a movement

(continued)

Table 3.1 (continued)

Progress in areas: information and communication technology and transport		Ideas and concepts in the organization of urban spaces	
			and city events. The only realization of the objectives of urban space is the Parc de la Gilette in Paris (1983–1992), (Tschumi 1996)
		1988	Asymptote Steel Cloud Project for Los Angeles West Coast Gateway—a deconstructed "City over City" (Brayer et al. 2005a, b, c, d, e, f, g)
The creation of the first social networking site Facebook, which allows for simultaneous and multidirectional flow of mass and individual content Collaborative mapping "Google Maps"	2004		The computational, parametric and generative design in urban planning: return to focusing on building typologies as landscape elements, borrowing forms from biological systems —forms of life (Hensel and Menges 2006)

The relationship between these two themes allow us to see the impact of different forms of communication and transport development on the shape of urban spaces

phenomena of larger scale and at higher levels of organization: national, ethnic, religious, cultural groups. It applies to building of a collective identity of nations and even the inhabitants of the continent. Mass communication, as opposed to direct information, does not contribute to the creation of ties between a user and the space of work or residence.[3] On the contrary, mass communication is associated with the breakdown of traditional neighbourhood communities (Goban-Klas 2009a, b) (resignation from the common space), since it is seen as a competition for traditional direct contacts. The natural need for functioning in the community and the sense of community is being satisfied through participation in the mass media. Social integration is always present on many levels: the neighbourhood level and even the family level may lose importance in favour of belonging to a much broader group (Goban-Klas 2009a, b). Promoting a variety of behaviour patterns, bringing us closer to geographically distant places they eliminate the fear or anxiety associated with migration. It loosens the human attachment to a known environment and increases his or her mobility.

A historically growing percentage of the population in the circulation of mass information is shown by the curve of the social range of information (Diagram 3.1). Access to information spread with time far and wide. The curve of the social range of information takes into account four main stages of development of mass

[3]An exception are multimillion metropolises where operate issuers of local mass media: regional television programs, radio channels, internet portals and traditional press.

Fig. 3.5 Panoramic views of the metropolis: the effect of growth of centres in the form of a dynamic panorama of the city can also be seen today. This tendency is often dependant on prestigious aspects, the motivation is to build a proper image, less often the need for direct business. Miami—a spectacular waterfront, favourable weather, positive associations attract investors. *Photo* A. Bonenberg

communication processes: the elite stage, the stage of popularization, the stage of specialization and the stage of multimedia (Mrozowski 2001). While the elite stage means communication within the highest social classes, the stage of multimedia means unlimited access to the mass communication.

Comparison of the curve with typical compositions of cities in different historical periods may suggest the occurrence of the following relationships: with increasing access to the mass media, we can observe a gradual policentralization and loosening of the city structure. At the same time, architecture loses its communicative features: an individualized detail, the codification by means of symbols and allegories once expressed in sculptures and frescoes.

There is no doubt that the issue of access to information is only one of many factors influencing the evolution of urban and architectural space. Apart from it, the primary importance had the population growth, specialization of functions, politics,

Fig. 3.6 Siena, Italy. In the skyline of the city is visible the spatial dominant feature: the tower of Palazzo Pubblico, the seat of the municipal authorities. Its architecture and scale are also an important element of the square frontage and indicate the important role that this object used to have. Architectural detail clarifies information concerning authorities, showing crests of ruling families in the elevation. *Photo* A. Bonenberg

cultural borrowings and many more. We should, however, pay attention to the consolidating importance of access to information in historic cities. Architecture was an important carrier of information about the place, about the prevailing social and economic relations. We could say that what was previously communicated by the architectural forms and their hierarchy and composition in the city, now may be transmitted by other means. The primary purpose of the comparison made is the observation of transformation directions and the diagnosis of contemporary spatial trends. Here crucial are the characteristics of spatial urbanization occurring in the information society era, that is in the stage of mass communication media.

3.1 Elite Stage—Compact and Clear City Composition

In the first elite stage of the information availability, participation of the mass audience was negligible. The time frames are here smooth due to the uneven development of universal education in Europe—which was a condition for receipt of mass media in printed form (sixteenth–seventeenth centuries). Earlier among the barriers to the circulation of information were illiteracy of a significant part of the population, and the overall access to global and local information took place in the public spaces of the city. The social, political order expressed by the urban

Fig. 3.7 Aylesbury Estate in London is a residential area, designed by Hans Peter Trentom, implementation of which lasted from 1963 to 1977. It includes 2,700 apartments, a population of 10,000 social housing built in line with the Charter of Athens. Today, it is one of the poorest districts of London. The degradation of the area progressed despite of differentiated building scale, projects of common green space and a significant financing of these objects. *Photo* A. Bonenberg

composition of the city and the architecture of buildings was understandable for the recipient. Legibility of urban space was based on the basic, common to all people level of knowledge, that is on associations and emotions: impressive, high buildings were important (churches, town halls), detail on the facades of tenements indicated the origin of the fortunes of rich merchants. The expression of artistic, decorative, sculptural elements in architecture was literal, and the scale of individual objects in the urban tissue corresponded to their rank.

Among the public spaces on special attention deserved such *public areas as stages for ritual and interaction.*[4] They were deliberately designed with a particularly significant activities associated with culture and rituals (Fig. 3.6).

[4]Places of particular importance: squares, markets in opposition to streets, ports and other strictly utilitarian space have additional cultural significance (Kostof 1992a, b).

Fig. 3.8 High Line Park in New York City, the project presented at the Venice Biennale in 2006, authors of the project: Diller Scofidio + Renfro Photo. Implementation of the park in 2010. The park has an elongated shape, suitable for jogging. *Drawn* A. Bonenberg

Since the impact on the associations and emotions is the most basic way to influence the community, the historical urban tissue is still so positively received by the general public. Referring to simple reflexes and associations, it is the most communicative way of shaping the city. Popularity of historic city centres as prestigious areas of trade, entertainment, tourism centres fully supports this thesis. These spaces are also conducive to traditional direct social integration.

Fig. 3.9 Mile End Park, London. Over the road, there has been suspended a wide bridge connecting the two parts of the park. The bridge is covered with lush greenery, a person passing through it does not feel leaving the park space. A spectacular design, which is the flagship of the site, was designed by CZWG Architects and opened in 2000. *Photo* A. Bonenberg

3.2 Popularization Stage—Polycentric Composition

At the stage of popularization, mass communication and access to information began to be diversified and better organized. It was linked in the eighteenth century with the development of technology: telecommunications, printing and transport networks, and previously with better access to education. There have appeared themed magazines, catalogues, books and newspapers. Access to information, though in a small degree, ceased to be a direct consequence of participation in public or social life: it independently reaches an increasing audience by mail.

Demographic and spatial development of European cities resulted in a partial decoupling of the system and the emergence of competitive local focal points. Communities focused around local squares, gardens and parks—access to local information was associated with entering the space of the nearest neighbourhood. At this stage, public spaces and spaces of social integration have become even more diverse and hierarchical. An important period associated with the general availability of the information in the urban space is the nineteenth century, when social organizations educating the poorest layers took the floor—funding reading rooms and libraries (such as Public Education Society and the Society for Public Reading —in Greater Poland).

Fig. 3.10 Intervention in the existing urban space: New Museum of Contemporary Art in New York, authors SANAA (Kazuyo Selima, Ryue Nishizawa). Foundation of the contemporary art museum was a part of the revitalization plan of surroundings of Bowery, the artistic neighbourhood of Manhattan, New York. The ground floor of the museum is an attractive common space, while the typical, sterile, closed shape of the object complements the frontage, at the same time becoming the spatial dominant feature of the area. *Photo* A. Bonenberg

3.3 Specialization Stage in the Industrial Cities of the Twentieth Century—Specialized System Versus Loss of Legibility of Urban Structures

Specialization stage of mass communication specifies the period of growth of public awareness in the field of civil rights, education and facilitating of initiating mass social movements. Education of population, increasing wealth, the ability to dispose the leisure time, awareness of the right to leisure, recreation influenced the layout of urban space. Naturally, therefore, the stage of mass communication specialization combines with the concepts of industrial cities and segregation of functions within them. As a result, there are created areas and districts of the city that are functionally homogeneous—specialized. There are separate industrial areas, recreational areas, residence areas and administrative service centres.

In the administrative service centres, of great importance is communication associated with the functioning of trade and services as well as public administration. The organization in the specialization stage of mass communication is based on simple tools of telecommunications (telephone, telefax)—so the number of

Fig. 3.11 Author's representation of the New Museum of Contemporary Art in New York. *Drawn* A. Bonenberg

remotely transmitted data between operators and administrations is small. Establishment of business activities in a good location (in the city centre) is a condition of access to customers, public administration and, in this sense, the condition of operation. This situation caused an intensification of city centres. Restrictions on information and communication technologies in the 90s of the twentieth century were a consolidating factor for new investments and to focus development.

An important aspect of social acceptance and readability of the city in the industrial era was the fact of their relationship with the intellectual culture of the European elite. City planning was entrusted to a well-known city planners and architects; it was associated with an advanced specializing of groups of decision makers and designers of new settlements and with the development of the theory of urban planning. The consequence of extensive theoretical, research and scientific base is the acceptance for spatial solutions whose justification has a scientific character, intellectual one or expressing ideological content. The source of information about the building destination was neither a readable form nor important object position in the urban tissue. A condition for reading the intellectual layers is knowledge—for people who do not have it, the space loses its communicative features (so well developed in the elite stage of mass communication). As a result,

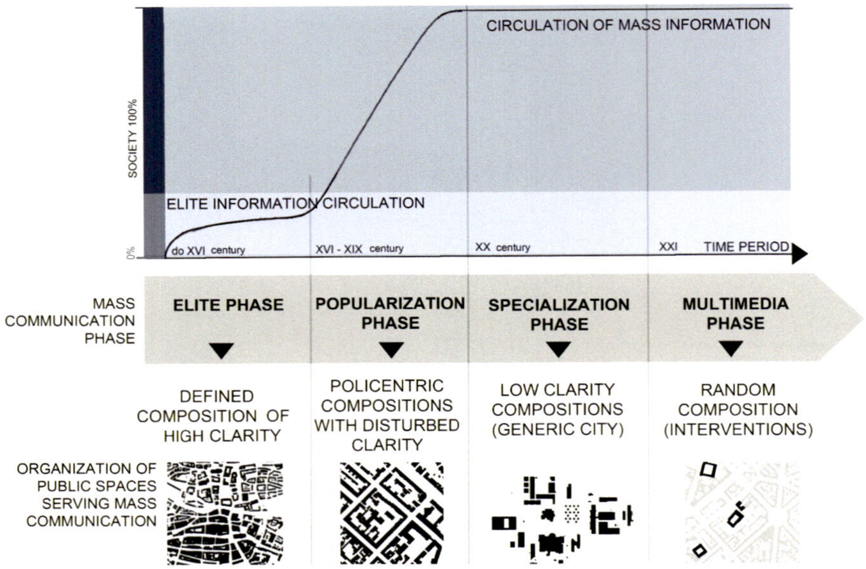

Diagram 3.1 A diagram illustrating the relationship between clarity of space and the stages of mass communication (Bonenberg 2013)

urban structures lost their legibility for a wide range of recipients. The result of these feelings in environments of architects and planners was the criticism of assumptions of the Charter of Athens in the 70s and 80s of the twentieth century (Fig. 3.7).

3.4 Multimedia Stage of Mass Communication—Urban Spaces of Information Society

In the last stage of mass communication, the multimedia stage, we shall emphasize the fact of "detachment" of mass communication and communication associated with the economic development from a real space. The Internet, developed infrastructure of data transmission, technical infrastructure, hardware devices cause that a significant amount of information can reach the recipient outside the urban space. Transmission of information ceases, therefore, to be at this stage, a centre-creative element, which consolidates buildings, decreases the importance of real spaces for a social and political organization, because the organization is beginning to be possible through information and communication technologies. In a hypothetical situation, in which the only parameter to shape the composition of the city would be a flow of information, an urban composition of the city in multimedia stage can be described as a "cloud" of design intervention: a disordered collection of architectural objects. Realistically, however, the appearance of the newly designed assumptions may depend on factors that are not associated with access to information and mean an increase in the creative freedom:

1. it can be a representation of the abstract elements of an artistic meaning,
2. it can focus on relations with the context of nature,
3. it can provide a group of new features "imposed" on the historic urban space—utilizing it without significant integration into the existing urban texture,
4. it can be inspired by personal interests and beliefs of the inhabitants.

Recent multimedia stage of mass communication takes place under conditions of severe loosening of the urban structure, partly as a consequence of urban development in the industrial era, but also as a consequence of decentralization deepened by the general accessibility of information and communication technologies.

The development of a European city in the multimedia stage can be described as a "cloud" of interventions conducted in the existing loosened structure. It involves the activities of revitalization, requalification of the destination of areas, corrective actions, as a consequence of departing from traditional ways of producing goods and switching to the creative activity based on access to information.

The dominant issues in the newly designed assumptions of settlements and suburbs are broad ecological issues: knowledge about energy issues and environmental protection. These projects often expose the natural conditions of the location.

In highly urbanized Europe and America, interest in the use of historic buildings, investments in expensive works of modernization and revitalization of facilities being a testimony of the material culture of past generations is the maturity certificate of societies and access to knowledge, which allows us to appreciate and enjoy this heritage. Along with longing for architecture and urbanism at human scale, returns the desire to use the space subjected to adaptation or revitalization—as exceptional, unique and original. In this cultural change, we can trace a popularity of lofts and all kinds of custom living space.

The location of the intervention is determined by a lot of parameters and a number of variables. Much to such an extent that the map of newly designed or revitalized urban objects seems to be disorderly and random. Factors determining the location of the investment are the technical condition of the existing buildings, ownership issues the availability of undeveloped plots, social conditions, political interests. In addition, interventions can have a very differentiated financial and spatial scale, they can be a social, public or private enterprise. Such nature of the activities has also much in common with the concept of urban acupuncture, developed by the Brazilian urban planner Lerner (2003), which expresses a set of activities aimed at activating and improving the urban space in the smallest scale. Effects of interventions both if they relate to urban scale (planning of settlements but also the urban development of the city square), as well as architectural, have a common positive feature: new and old elements in the urban structure do not stand in opposition to each other: on the contrary, they may be based on good neighbourly relations and existing natural and cultural environment. What dominates here is a holistic approach and consent to the multithreaded approach to the composition of the city. Described attitude of acceptance and using resources is valuable. According to the author, as opposed to the concept of industrial cities, from the partially legitimate demands of The Athens Charter, urban spaces of the information society can be aimed at exposing of existing cultural and social resources.

Many of the great conceptual designs, for example the plan of Le Corbusier for Paris, were manifestos of the coming revolution—beginning of the new order. The tool to achieve this new order was to demolish the existing urban fabric. The revolution was the voice of dissent to the order of the past, the emotionally charged concept, full of rebellion against the low standards of living in the historic cities. The urban space design concepts for the information society are not based on the revolution. They are based on improving the standard, urban and architectural interventions. They are also based on diversity, giving users the choice—where and how to live. Some residents can pursue a fascination with the past, hence the numerous adaptations and revitalization of the historic building tissue. Other users can realize the closeness with nature—in projects aimed at broad ecological values. Not without significance is the economic balance in the world living in the shadow of the energy crisis and the awareness of the importance of energy efficient solutions for the environment. Comparing both assumptions, the industrial city treated a user of urban spaces in a more objectified way, not giving such a choice. This approach was, moreover, associated with the social order arising precisely from industrialization: the collective work in the industry or office work.

For designers, there is an open range of possibilities based on the foundation of ecology, respect for nature, references to traditional and historical architecture (increase in importance of vernacular movement) as well as the application of advanced computational design methods.[5] Within one city or metropolis, we can find and appreciate the multithreading of buildings: from architecture of advanced technology solutions of traditional concepts and even archaic ones. From the environments under full electronic control to the projects constructed with earth created with hands of their future residents. Thus, the structure of the city becomes multilayered and rich. A user of the urban space, a representative of the "creative class", understands this order and benefits from diversity. Access to it stimulates his creativity. Dynamics and ability to innovate are the greatest value in the information society. The multilayered city space can both emphasize it and become a result.

Use of the existing urban tissue involves the need of improvement of the quality of the surrounding public space. This may lead to the demolition or intensification

[5]Computational design allows the generation of architectural forms using digital optimization methods, in which the architect controls the form using a programming code.

of development. The aim is to achieve an acceptable high standard of the architectural and urban space. One of the best ways of improvement is the introduction of attractive places for recreation and greenery in an urban space.

An example of the urban life celebration through the urban greenery project is the High Line Park in New York City. The objective of the project is to adapt the unused industrial port railway that was preserved in Manhattan into an unusual park with an elongated shape: with a high-quality greenery, small architecture and artistic installations, thematically related to the environment. Historical elements of the structure of the city can become in the process of revitalization, a valuable space attracting and gathering members of the creative class.

In parallel with the New York project, in London, there was created a park of a much more natural character and with a much larger area. It was created on former industrial areas, demolished during the Second World War. Noteworthy is a pro-social nature of this urban investment—aimed at improvement of the living conditions in a multicultural, multiethnic and indigent district—Mile End Park.

Both investments constitute an intervention in the urban space in order to prevent its degradation, while creating and maintaining a public social space (Figs. 3.8 and 3.9).

Often the choice of location to live and work is at the same time determined by several items and very personal issues: personal feelings and preferences. Spatial interventions in the city have a character that is difficult to define, because the interest of the residents, entrepreneurs, investors in a given area may be a matter of many different factors:

1. cultural (fashion trends),
2. business and economic,
3. infrastructure (good communication),
4. functional (availability of essential services),
5. environment (presence of attractive elements of the natural or urban context),
6. political.

Single, scattered changes in the urban tissue have a limited radius of action and do not affect the degree of fragmentation of urban space.[6] A significant improvement in the perception of the city and its functional integration can be achieved through an appropriate coordination of investment. Such planning is an increasingly common practice. The aim is to determine the location of particular importance to create activation corridors or axes that could restore the impression of a space integrity.

The availability of new communication tools and the continued availability of mass media including social media give us the opportunity to initiate a continuous process of review and improvement of the designed and implemented components.

[6]The problem of fragmentation of the city is associated with the increasing trend of spatial divisions in cities: social, economic, cultural and political ones. The phenomenon is socially harmful because it highlights differences between the inhabitants—generating conflicts and, generally, increased crime (Graham and Marvin 2001).

Such an action can be a tool for coordinating the cloud system of intervention and serve choosing the urban development strategy of information society (Figs. 3.10 and 3.11).

References

Bonenberg, A. (2013) *Media, przestrzeń, architektura. Transformacje przestrzeni społeczeństwa informacyjnego*. Poznań: Wydawnictwo Wydział Architektury Politechniki Poznańskiej.
Brayer, M. A., Migayrou, F., & Nanjo, F. (2005a). *Archilab's urban experiments. Radical architecture, art and the city* (p. 119). London: Thames & Hudson.
Brayer, M. A., Migayrou, F., & Nanjo, F. (2005b). *Archilab's urban experiments. Radical architecture, art and the city* (p. 84). London: Thames & Hudson (119a).
Brayer, M. A., Migayrou, F., & Nanjo, F. (2005c). *Archilab's urban experiments. Radical architecture, art and the city* (p. 114). London: Thames & Hudson (119, 84a).
Brayer, M. A., Migayrou, F., & Nanjo, F. (2005d). *Archilab's urban experiments. Radical architecture, art and the city* (p. 73). London: Thames & Hudson (119, 84, 114a).
Brayer, M. A., Migayrou, F., & Nanjo, F. (2005e). *Archilab's urban experiments. Radical architecture, art and the city* (p. 186). London: Thames & Hudson (119, 84, 114, 73a).
Brayer, M. A., Migayrou, F., & Nanjo, F. (2005f). *Archilab's urban experiments. Radical architecture, art and the city* (p. 812). London: Thames & Hudson (119, 84, 114, 186a).
Brayer, M. A., Migayrou, F., & Nanjo, F. (2005g). *Archilab's urban experiments. Radical architecture, art and the city* (p. 151). London: Thames & Hudson (119, 84, 114, 186, 812a).
Gehl, J. (2009a). *Życie między budynkami, Użytkowanie przestrzeni publicznych* [Original title: Life between buildings: Using public space] (p. 9). Cracow: Wydawnictwo RAM.
Gehl, J. (2009b). *Życie między budynkami, Użytkowanie przestrzeni publicznych* [Original title: Life between buildings: Using public space] (p. 12). Cracow: Wydawnictwo RAM (9a).
Goban-Klas, T. (2009a). *Media i komunikowanie masowe* (p. 119). Warsaw: Wydawnictwo Naukowe PWN.
Goban-Klas, T. (2009b). *Media i komunikowanie masowe* (p. 121). Warsaw: Wydawnictwo Naukowe PWN (119a).
Graham, S., & Marvin, S. (2001). *Splintering urbanizm*. Padstow: TJ International Ltd.
Hensel, M., & Menges, A. (2006). *Morpho—ecologies* (p. 16). AA Agenda No. 4., London.
Jellicoe, G. (1961). *Motopia: A study in the evolution of urban landscape*. London: Studio Publications.
Kostof, S. (1992a). *The city assembled* (p. 194). London: Thames & Hudson.
Kostof, S. (1992b). *The city assembled* (p. 123). London: Thames & Hudson (194a).
Lerner, J. (2003). *Acupuntura urbana*. Rio de Janeiro: Editora Record.
Levine, N. (1996). *Frank Lloyd Wright* (p. 221). New Jersey: Princeton University Press.
Malisz, B. (1966). *Zarys teorii kształtowania układów osadniczych*. Warsaw: PWN.
Mrozowski, M. (2001). *Media masowe, władza rozrywka i biznes*. Warsaw: Oficyna Wydawnicza ASPRA-JR.
Nebois, T. (Ed.). (2003a). *Architectural theory from the renessance to the present* (Vol. 2, p. 680). Cologne: Taschen.
Nebois, T. (Ed.). (2003b). *Architectural theory from the renessance to the present* (Vol. 2, p. 704). Cologne: Taschen (680a).
Tschumi, B. (1996). *Architecture and disjunction* (p. 141). Cambridge: MIT Press.
Wujek, J. (1986). *Mity i utopie architektury XX wieku* (p. 65). Warsaw: Wydawnictwo Arkady.

Chapter 4
Strategies for the Development of Public Spaces in the Era of Mass Communication

Abstract In order to obtain high efficiency of spatial resources, it is necessary to have vision, planning and adapted legislation. Space planning, optimized management of infrastructure and resources based on information and communication technologies combined with activation of urban space through social media makes the relationship between the space and the user to become dynamic.

Multimedia stage of mass communication can be understood as entering the relationships that are characteristic of the information society. The economic situation on which depends the spatial development becomes particularly unstable in terms of global competition and the global economy. The ongoing economic crisis shows that the inhabitants of the cities, in order to significantly improve their living conditions without incurring large costs must, in addition to the basic operations to maintain urban tissue and infrastructure, focus on the processes taking place in the city. The priority should be to optimize the management of infrastructure and spatial resources based on information and communication technologies. Optimization allows efficient use of space, resources, infrastructure, means of transport, public meeting space, and possibly the working space—provided, however, that the proposed spaces (both architectural and urban) will be flexible and multifunctional. In times of the economic crisis, the only contemporary advantage over the period of financial prosperity in 80s and 90s, and a chance to improve the urban conditions is communication and the ability to optimize, which are specific to the systems we use. Social aspect of information and communication technology enables activation of public and private spaces, verification of its quality and optimized management through augmented reality. The development does not necessarily mean the continuation of mass urbanization known from the past decades, only better use of existing resources, including point interventions in the urban tissue. It is also the answer to the demands of a sustainable development: less urbanization means more natural areas, concentrated building saves technical infrastructure, environmental and economic costs of bringing the media. Better standard of living can mitigate town sprawl trends and the problem of crisis of metropolitan centres. Access to information and communication technologies can

A. Bonenberg, *Cityscape in the Era of Information and Communication Technologies*, The Urban Book Series,
https://doi.org/10.1007/978-3-319-69542-6_4

affect a city at the level of planning and design, space and resources management and the construction and activation of the community based on the new media.

The challenge is, therefore, to build IT tools to improve the use of urban space. A variety of experimental proposals are created in the leading scientific centres. The interest in programs that can potentially optimize and enhance the perception of the environment is driven by a race of innovation in IT and telecommunications industries. For the time being, scientists and artists make up mostly theoretical or pilot scenarios from the borderline of functional art and computing, and interesting works are carried out by Media Lab MIT, Interaction Design Institute Ivrea[1] (Italy), Department of Computer Science Rutgers University, Waag Society in Amsterdam, Royal College of Art in London. The basis for most projects associated with the use of space is authentication using public collaborative mapping representing the aggregation of existing web mapping (Google Maps, Wiki Maps, Zumi) and the content created by users. Digital maps can act in cooperation with GPS receivers and be extended with the content of individual users. The factual content of the map (texts, images, films, videos) depends, to a large extent, just on them. In areas where the use of maps is popular, the amount of available information concerning the space has been huge at this moment. Thus, the primary data collection tools exist—now, we must find their application in optimizing the use of urban and architectural spaces. The aim is to enable societies to use massive amounts of data that produce.

One of the proposals of technical solutions comes from Samir et al. (2005), a pioneer in the field of mobile data. Observing the increasing use of various types of sensors, which are capable of networking, he developed methods of their use for data transmission and processing. The most common devices include: road sensors, environmental sensors (indicating the amount of rainfall, pollution, light intensity), mobile sensors equipped with a GPS system. All of these devices not only provide information, but also often have the ability to store it. A virtual physical space is formed in which most information is stored where it was created and not, as is currently the case, on remote servers. This space is called Dataspace (Franklin et al. 2005) meaning physical space underpinned by digital information.

Ideas based on the use of mobile devices as tools enabling to interact with were proposed by scientists from Ivrea. They believed that the smart phone, computer or tablet can function as a remote control triggering interface in the environment, in the public space. Originators draw attention to the fact that companies such as J.C. Decaux or Viacom Outdoors support tens of millions of LCD screens in public space: signage, advertising, public transport timetables. Availability of network resources through such elements could be a source of reception/transmission of information. Of the forty different scenarios of use of a mobile phone developed

[1]Interaction Design Institute Ivrea (Ivrea Interaction, IDII) was a scientific and academic project for planning interaction (Interaction Design). The activity was aimed at creating new business strategies, use of technology, innovative combining of issues from the borderline of science and social issues. The main co-organizer and sponsor was Telecom Italia. Ivrea offered the second degree of studies of *Interaction Design*, after five years of activity connected with the Domus Academy.

in Ivrea, an interesting concept is zones created by small architectural elements (such as benches), enabling the conversation on our own mobile phone but using external speakers with an integrated seating. An interesting example of the application of such a solution may be a thesis (Lee 2006) from which resulted the Sonic Graffiti—a system allowing for leaving in the urban space a composition of sounds, signals which can be received by another person. In this way, a piece of music or other signal can be assigned to a specific location, which will have an impact on the perception of the passage of urban space and its identification.

In the architectural scale, co-working offices are an example of optimizing the use of space.[2] Working space is occupied temporarily, depending on your needs—its use depends on good organization between users. Availability of meeting rooms, conference space and even desks is organized temporarily, on the basis of negotiations. In the traditional model, each user would need his own desk, work alone or in a fixed circle, incurring much higher costs, consuming a lot more resources, space and basic media: electricity, heating and many others. Variability of use is a particularly good, cost-effective solution, building simultaneously interdisciplinary conflicts: inspiring and economically encouraging.

4.1 Social Media

A particular type of mass communication, which can be helpful in the study of urban space are social media. Social media, functioning as a multidimensional forums for exchange of views and ratings, optimize contacts in groups, function as an effective channel of communication. They have participated in the creation and promotion of the concepts, creative ideas and decisions, including those leading to important spatial transformations. The network of contacts and linkages that exist in the virtual world, where the adhesive are common interests, work, system of values, social environment of users, will develop and their impact on the spatial environment will grow.[3] Visions of urban spaces that emerge from social media are very diverse. Certainly, they take into account the tastes and needs of the information society—the users of this medium.

---

[2]*Co-working* is a style of hiring the job space, which consists in using a common working environment, the most often office space for many people conducting independent operations. In contrast to the traditional model, these persons are not employed by one organization. Using co-working offices is characteristic of microentrepreneurs, small businesses and start-ups.

[3]Many short-term municipal actions, for example associated with the planting of trees in undeveloped areas, the organization of municipal games, is based on instant communication.

The impact of social media on the decisions concerning the development of the space can be divided into two basic groups:

1. access, commenting, communication of key information for planning decisions (public consultations),
2. organization, activation of space.

Giving the opportunity to stay in touch with a wide range of users, social media allow us to influence, through marketing tools (social media marketing) on the situation in the city space. They become a tool for the creation and distribution of trends, also in relation to specific areas—places of meeting in real space. Social media allow to observe the dynamics of lifestyle trends, preferences for choice of places of residence, leisure and meetings. There is a relationship between the transmission of information in social media and the degree of development and the quality of urban space and architecture.

An important feature of communication in social media is its multifunctionality and the lack of established, traditional hierarchy of the flow of content. Social media democratize information transfer abolishing the monopoly of press and television by ensuring the access for all and dramatically lowering the cost of mass transmission of information. Anyone with access to Internet can potentially become the issuer of the content on a massive scale. Disturbance of the traditional hierarchy: mass media—recipient radically changes the situation of an entity that receives a real possibility of transmission of his or her own message and its content. The direction of the flow of information "all to all" means the opportunity to organize and initiate activities both locally and globally.

Social media sites differ from one another due to the preferred form of expression and the kind of interaction between users. They can be systematized because of their fundamental tasks:

1. publication as a blog or Web log: text, images, videos, which retain the chronology of entries and have characteristics of a diary. Popular providers of this service include Blogger, Tumblr, Wikipedia,
2. publication of videos, photos in the portals: YouTube, Flickr, SlideShare,
3. tools for direct verbal communication and video conferencing: Google Talk, Skype,
4. services to allow association of friends and sharing information in their group. Facebook, Google plus,
5. a tool to send short text comments, images or films in real time: Tweeter,
6. entertainment in the group: online games,
7. tools associated with geographic location: tell about the place, tell where you are and whether you like it there (Diagram 4.1).

The main argument in determining the usefulness of the tool, which are social media, in identifying and deciding on the shape of the urban space is the fact that they are also an important part of culture creation. In this sense, they not only give

Diagram 4.1 Possibilities of using social media for different purposes of creating, modelling and education. Created by A. Bonenberg

the opportunity of passive observation of feelings and evaluation of space, but with their help we can shape these views. Social media make it possible to:

1. have insight into the subjective and spontaneous feelings of space users,
2. establish interaction and emotionally bond the user with the place,
3. influence feelings about the space using techniques of territorial marketing,
4. determine the user's preferences—in order to predict his or her needs,
5. monitor opinions in time, because perception and evaluation of space is constantly changing, and exchange of comments and evaluations is carried in the real time, concerning also short-lived phenomena in the urban space,
6. use various forms of expression,
7. easily obtain statistical data, the number of comments: "I like it", "I recommend it".

4.2 Public Consultation

Social media can be a tool of public consultation in urban planning, which will result in co-creation and co-determination of the urban space. They are used in the economic, social and functional activation—through territorial marketing techniques. Both goals (participation and activation) provide an impetus to improve the quality of urban space through architecture, new interior planning and design. In such a case, they are created thinking about the target group: participants in social media.

Social media as a tool of social participation enable efficient management of shared resources (*sharing*) of infrastructure, transport, public spaces, in accordance with the principles of urban development strategy of information society. They also support the free access to information and promote transparency of decisions taken centrally. Facilitating methods of participation: *social media crowd sourcing* is a cheap and efficient acquisition of information about the place.

The main and serious drawback of social media is the lack of an effective tool for organizing and sorting of information. A huge amount of the collected data is sometimes difficult for correct interpretation. In addition, on implementation of the plans it was noted that participation was attended by just a few leaders—such as in projects of participation in the real world. This means that a group of people deeply engaging in projects is similar to that involved in social participation carried out in the traditional way.

4.3 Urban Space Activation

The purpose of marketing activities through social media is making people interested in a particular place in space: generating the trend and fashion to remain in the specific area. Better recognisability and increased confidence encourage new members to reside in this space. Such actions may be commercial in nature and they are widely used in marketing: shopping malls, clubs, pubs, restaurants and shops. Non-profit activities are generally carried out by local authorities, and their content is a widely understood popularization of the advantages of the city, municipality, province and its achievements, cultural and natural heritage.

The effect of media message on the public space of the city has different consequences and follows a characteristic pattern. The consequences can be described as functional, social and visual. The impact of the long-term media message on the public space of the city can be described as follows:

1. social media, as a result of information transmission and discussion between users about the site, attract people by raising their interest in a particular portion of the public space,
2. interest results in physical presence of a large number of people,

3. the presence of people results in economic and social activation of that public space,
4. the economic and social activation impacts on the quality of the space. Visual quality depends on the preferences and tastes of the people who use the space.

Creating a permanent trend stimulating positive action is possible and it brings tangible economic and social benefits. In addition to supporting important institutions of culture and education, social media allow activation of urban spaces in the form of short, dynamic actions. While in the case of permanent institutions, social media can be used as an assisting tool (actions of an institution are not dependent on the actions of social media), then in the case of short-term events efficient flow of information is the basis for the success of the event. As an example, we may point at shops or pop-up bars, which depend on the efficiency of information flow. In the Australian example of the three-day store, the intriguing interior emphasized a temporality of the event, ecology, flexibility.

The group of short-term events may include, in addition to commercial activities, cultural and artistic events. One could say that the impact of information transmission on urban space depends on the length of actions undertaken as its consequence. Constant activity through social media, which is associated with the action for the promotion of space, may result in an effective, long-lasting activation of space. Short-term message through the media, which is associated with a specific event, contributes primarily to consolidation of the positive image of the place.

4.4 Design Verification

With mass communication, we can influence the usage of space and we can also verify and evaluate on a regular basis its functioning. Verification and evaluation of the space, services, urban facilities can be made based on social media, and by designing spaces that are flexible, susceptible to change, verification can result in adaptation—and then another, relatively low in cost spatial change—subject to further verification. Important here is the continuity of the cycle, giving the effect of continuous improvement, adaptation and updating of solutions applied in the space through consultation, activation and verification. Users of the space receive the basic benefit: constant contact with the space of a good standard rather than a model in which the place is brought to a complete degradation in order to completely rebuild it (Diagram 4.2).

In order to obtain high efficiency of spatial resources, it is necessary to have vision, planning and adapted legislation. Spatial decisions can be consulted and discussed, and there is access to multiple experiences gained from other places: exchange of best practice models, mutual cooperation and promotion. Space planning, optimized management of infrastructure and resources based on information and communication technologies and activation of urban space through

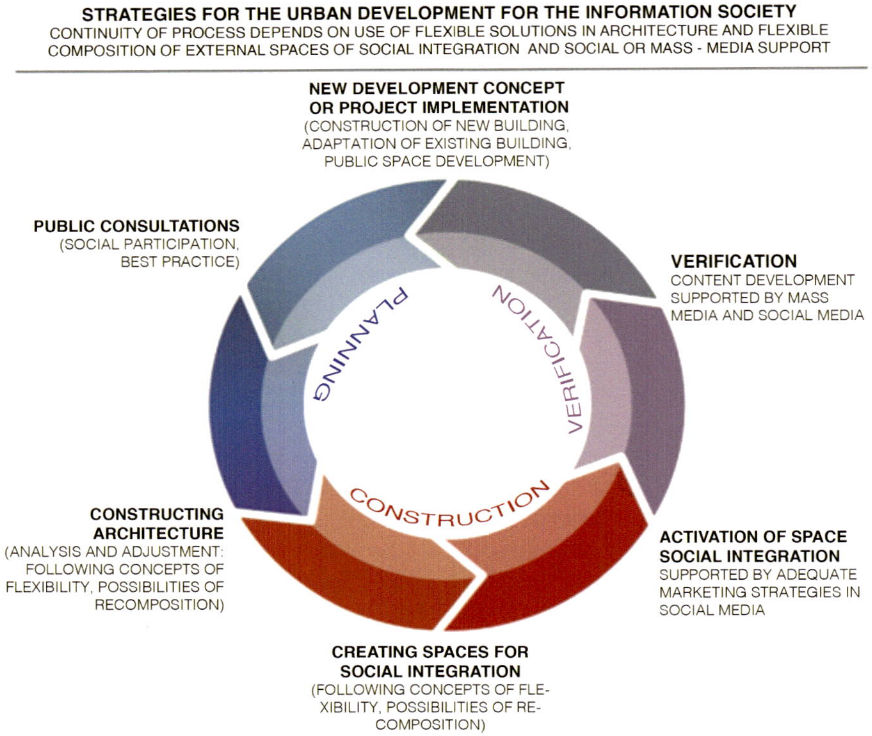

Diagram 4.2 Strategies for the urban development of the information society. Local communities take actions when they are well integrated: they easily to define their own needs and they are able to turn them into concrete actions. The organization and communication within the local community can be simplified by the social media. Created by A. Bonenberg

social media make the relationship between the space and the user to become dynamic and, in consequence, it will cause a variability in functional and compositional architecture.

References

Franklin, M. J., Halevy, A. Y., & Maier, D. (2005). From databases to dataspaces: A new abstraction for information management. *SIGMOD Record, 34*(4), 27–33.

Lee, C. L. (2006). *Sonic Grafitti* (Master Thesis). Interaction Design Institute Ivrea.

Samir, G., Imielinski, T., & Ozbay, K. (2005). Ascertaining the viability of WiFi based vehicle-to-vehicle network for traffic information dissemination. In *Proceedings of the 7th Annual IEEE Intelligent Transportation Systems Conference (ITSC).*

Chapter 5
Cityscape: Dominant Functions and the Role in the City Structure

Abstract Understanding objectives of the contemporary cityscape design as a harmonious, sustainable development on the basis of the past structures, cities for information society should be based on diversification and multithreading. This approach creates potential for fixing current problems: crisis of city centres, disintegration and economic, social and cultural fragmentation—as well as the urban sprawl phenomenon. The development does not have to signify a continuation of mass urbanization known from the past decades, only a better utilization of available and owned resources.

To build and establish a society based on knowledge requires a development or strengthening of some of its sectors such as education, continual teaching to expand gained knowledge and transfer of technologies. Facilities related to the above-mentioned functions constitute significant reference points in city composition, and it may be expected that their role will grow in both the social dimension as well as spatial dimension. Table 5.1 allows for an abbreviated depiction of consequences, which will be a part of city space as a result of the development of information society, as much as it can be defined at this time.

Activities in the urban space are the result of processes contained in Table 5.1. New investments represent an opportunity and excuse for positive spatial changes, better definition and improvement of readability of the urban space. The current modernization and creation of new facilities should take into account the occurring social changes, namely:

1. meet the conditions of interaction, communication, exchange of views, both in the physical spaces of the object and by activity in the space used by a virtual organization or institution,
2. should emphasize the importance of social interaction, redefine public spaces where it is possible to implement joint actions,
3. should have a positive impact on the readability of the city, contribute to the possibility of identifying its fragment,

A. Bonenberg, *Cityscape in the Era of Information and Communication Technologies*, The Urban Book Series,
https://doi.org/10.1007/978-3-319-69542-6_5

Table 5.1 Spatial phenomena taking place in the city resulting from the development of information society; devised by the author based on "Raport Kongresu Informatyki Polskiej", (*"Congress of Polish Information Technology Report"*), Poznań, 1997

Effects associated with the development of the information society	Spatial consequences in the city scale
• The creation of the global information economy, in which access to information is a key factor of the performance of industry and services. The challenge is to face the global competition in the conditions of internationalization of the economy, increase in the competitiveness between internal and external labour market. It comes to lowering of labour intensity of production processes through automation and computerization and the displacement of some branches of the economy by others	• Relationships in the scale of cities and regions have a character of networks in which global metropolises and big cities become dominant points. In the scale of metropolis and cities, there is a significant dynamics of change in their functions. Due to changes in production technologies, diminishing of industrial areas there can be observed changes in the city structure. There occur: loss of function, intensification of building in some areas and spatial expansion (town sprawl)
• Creativity and innovation becomes a driving force of economic growth. The primary objective of entrepreneurs and regional authorities is to multiply human capital by attracting creative people, highly qualified, capable of innovative, customized solutions. The aspirations for high-quality education	• The desire to shape the high-quality life space based on the local natural and cultural resources in order to enhance and individualize the conditions of residence. Opportunities relate particularly to city centres, their exceptional visual quality and attractiveness of "urban narrative"
• Acceptance of a variety of styles, rhythms of life, which result from the popularization of irregular working hours, increasing the possibility of performing multiple occupations over the Internet	• Providing public goods in flexible hours, attention to nocturnal exposure and usability of urban spaces
• The aspirations for high-quality education. The need to develop modern educational centres, providing universal education, with an emphasis on technical and computing fields Development of lifelong learning which is a consequence of technological advances, changes in production methods and multiple outdating of professional skills in the life cycle of a human Development of methods for distance learning	• The construction and expansion of the network of educational, research and implementation institutions. Extending them with functions associated with implementation of the achievements of science in industrial production: supporting the creation of science and technology parks
• Decentralization of location of work places as a consequence of changes in the organization of the services and industry sector. Generalization of "teleworking" performed at the place of residence. Increasing integration of work and home life, working and leisure time, which are important for social and psychological ties and relations as well as of the models of life	• Designing of objects with a flexible interior design, easy to adapt to the changing functions. Construction and adaptation of living spaces, providing adequate conditions for working

(continued)

Table 5.1 (continued)

Effects associated with the development of the information society	Spatial consequences in the city scale
• Changes associated with culture and customs, whose important element is a continuous access to mass media, including social media. The important role of mass media and uninhibited flow of the content of communication in shaping culture	• Constant access to information in the urban space, the opportunity to influence social behaviour, manner and intensity of use of urban space (popularization of places via social media) • The ability to easily conduct a public consultation relating to urban and architectural planning • Multimedia screens, multimedia facades as an integral part of the architecture. The popularization of multimedia commercial screens in the urban space and its negative impact on the space
• Working in office conditions, improvement of the physical working conditions, access to health care has its share in the extended life expectancy	• Investment in infrastructure, housing for the elderly persons

4. may have a flexible and variable composition of the interior spaces to meet the assumption of verification and correction of the original objectives, as described in the strategy of the urban development of the information society,
5. should give the possibility to create personalized, diverse, inspiring spaces,
6. be constructed in accordance with the principles of a sustainable development.

Examples of this type of successful architectural projects exist in many cities; we can mention here the new facilities built in the USA: the science museum in San Francisco by Renzo Piano opened in 2010, Stata Centre Massachusetts Institute of Technology designed by Frank Gehry and many others. Facilities dedicated to creating and spreading knowledge, such as those mentioned above, are increasingly used as the public meeting places, creating forums of social communication. They also play an important role in both physical and virtual space. Museums, libraries, centres of learning and culture bring together members, fans and guests thanks to the mass media, and above all, thanks to social media. Media transfer proclaims availability, openness of these organizations for all stakeholders, which builds a society of knowledge (Figs. 5.1, 5.2 and 5.3).

The choice of the location of these functions should contribute to:

1. Restoring the importance of a city centre: if access to global information becomes less important due to extensive data transmission technologies, access to local and direct information can be a major driving force shaping the social integration in the city. Promoting the "urban narrative"—a culture of life in the city as a factor counteracting to the town sprawl phenomenon.
2. Preventing the disintegration of urban space. A significant improvement in the perception of the city, its functional integration can be achieved through an

Fig. 5.1 Opened in 2010, the science museum in San Francisco by Renzo Piano, erected in the heart of Golden Gate Park. The building is covered with a hilly roof planted with low vegetation, typical of the area, and in the background one can see an analogous outline of hills, so characteristic for the cityscape. Aimed in its concept at the promotion of science, the building seems to be composed of a greenery, air and water. The space for integration, lectures, demonstrations has been situated in the heart of the facility—museum focused on education and raising awareness about sustainable development, ecosystems and activities devoted to their preservation. *Photo* A. Bonenberg, *Fig.* A. Bonenberg

Fig. 5.2 Renzo Piano Building Workshop, the science museum in San Francisco

Fig. 5.3 Frank Gehry MIT Stata Centre in Cambridge: the building of the Massachusetts Institute of Technology has a lobby with a colourful interior, cafeteria, exhibition area of student works on the ground floor, extended with a space of mezzanine. Collapses in directions of outer walls create intimate alleys, illuminated with natural light. Drawing A. Bonenberg

appropriate coordination of investment. Creation of activation corridors or axes that can restore the integrity of the space.

3. Creation of newly built areas by method of neighbourhoods: the principle for creation of districts and neighbourhoods with a rich service and commercial programme, public area of sport and recreation. The more residents are freed from having to live close to work, the more the quality of space around the house will determine the choice of place of residence.

Fig. 5.4 Jean Nouvel Ateliers, Kilometro Rosso, Begamo: a new building typology focused on the innovative use of knowledge in the industry. *Photo* A. Bonenberg, *Fig.* A. Bonenberg

4. Making use of the availability of new communication tools, the continued availability of mass media including social media to initiate a continuous process of review and improvement of the designed and implemented components of the urban tissue (coordination of the cloud system of intervention. Optimizing the management of infrastructure and spatial resources based on information and communication technologies.
5. Revitalization of existing space to meet the conditions of flexibility and in the case of unsuccessful verification to be able to introduce the required spatial changes.
6. Maintenance of the reserve areas to invest for the industry in the context of the voices of the need to reindustrialize Europe. The issue is important in the context of requalification of industrial areas (historic areas associated with heavy industry).
7. Development of new spatial solutions to order building typologies of objects for technology transfer and education. An example of such an arrangement of space is a project by Jean Nouvel Ateliers, Kilometro Rosso in Begamo (Fig. 5.4).

The discussed above standards will allow cities to develop in a more balanced, economic and just manner from the point of view of access to living, working and relaxation places. Development does not have to signify a continuation of mass urbanization known from the past decades, only a better utilization of available and owned resources. A city of the information society may ensure a more socially fair access to comfortable, common space and public usage facilities, the proof of which are the realizations and design projects of city districts created by Renzo Piano. Building or creating on the already existing groundwork, cities for information society based on diversification and multithreading is an antidote for problems of a modern city: crisis of city centres, disintegration and economic, social and cultural fragmentation—as well as the *urban sprawl* phenomenon.

Chapter 6
Designing the Living and Working Space of the Creative Class

Abstract Design of spaces for the creative class is based on supporting creative actions through appropriate formation of space. Personal physio-psychological state, ability to concentrate, ability to think creatively, general well-being, comfort, inspiration and mood are important factors of the creative process. The combination of spatial and psychological studies is the basic necessity in creating spaces for the creative class.

Contemplating the influence of information and communication technologies on architectural and urban space, it is necessary to take under consideration social and economic changes. General and open access to information seems to have its reflection in goals, needs of increasingly growing group of people—creative class. In the industrial era, the rhythm of the cities was set by the daily operation hours of factories and offices: the division between the place of work and home was distinct. In environments where the status and economic development are based on the branches of creative industry, creation of knowledge and processing of information, this rhythm is less regulated. This is because frequently it is based on the operation of the units of "freelancers" as well as small and medium companies, and the borderline between place of work and home is becoming obliterated and thus less distinct. Information and communication technologies allow to work from home: creating items, consulting projects and selling them regardless of geographical location. Economic success depends on intellectual work and the ability of finding innovative solutions.

The socio-economic concept of the creative class was created in the USA as a result of studies within the Standard Occupational Classification System (SOC) framework. It is estimated that in the USA, creative class constitutes 30% of the entire workforce. The secondary segment introduces the "superactive core" and "creative professionals". Their work does not have to be subject to defined time-frames and is an individual process. The customary effect of change is an increase in the acceptance level of diversified lifestyles, as well as non-regulated, flexible work hours. Freedom involved with task evaluation of work results, free choice of work performance location (generally through the use of the internet)

A. Bonenberg, *Cityscape in the Era of Information and Communication Technologies*, The Urban Book Series,
https://doi.org/10.1007/978-3-319-69542-6_6

accustoms expression of one's own concepts and individualism.[1] As a group, creative class is differentiated by rejecting many conventional behaviours and the tendency to create less formal relations free of stiff rules. The effect of this is abandoning formal attire but also alternative arrangement of living and working space. Personal preferences, individual style and character often become expressed by individually created approach towards shaping ones living space. Creative employees often operate in a network of other creative co-workers. On a principle, their interrelations are democratic and a great deal less formal than those existing in corporate hierarchy. In this sense, the access to information and communication technologies weakens the stiff and rigid formal hierarchy, which is a reminiscence and foundation of the industrial society.

6.1 The Creative Process and Spatial Environment

An appropriate psychological approach significantly impacted by environmental situation is the primary requirement for creative position. Environmental situation favourable to creativity may be understood as a set of factors motivating goal achievement (Nęcki 1994). Therefore, stimuli flowing from the surroundings influence affectivity of the creatively characteristic performed work. Personal well-being, comfort, concentration, inspiration and mood are important influential factors course of creating that is on the creative process. Spatial surroundings definitely have an impact on the main physio-psychological state of the user; thus, from this point of view, the application of spatial studies is the basic necessity of creating space for the creative class.

Numerously published work space projects of the leading IT companies (Google and Facebook), the potential of which depends on the creativity of its employees as well as innovation of applied solutions, prove the thesis that multifarious, personalized and informally arranged spaces—inspire and incite creative process.

It should be kept in mind that: "*physical architectural form, creating a meso environment for an individual, is a source of stimuli for an individual invoking defined reactions as well as its performance aspect ensuing from the fact that architectural situation is an area in which a person develops one's activity (…) Thus, when defining various manifestations of human behaviour in architectural space, not only the fact that they are a reaction to that situation should be considered but also that some behavioural manifestations are a solution to this situation*" (Bańka 1997).

[1]The term "creative class" was created and defined by an American economist and sociologist Richard Florida, University of Toronto, by so doing he defined a class, which was the main power force of economic development of post-industrial cities (Florida 2010).

Therefore, the basic issue is the subject of supporting different behaviours and qualities of the user, which allow him to create value. Space should provide stimuli for defined, desired behaviour and support basic devising methods:

- free association method,
- method of observing similarities,
- associating remote facts and ideas technique,
- easiness in understanding and using a metaphor (metaphoric thinking),
- completing far-reaching transformations,
- use of creative imagination (Fig. 6.1 and Table 6.1).[2]

The examples of studies' application pertain to designing residential and community spaces. Each of the quoted examples of projects and their execution represents one of the creativity supporting methods:

1. Application of flexible solutions in architecture: this to facilitate its adaptation to the dynamic changes in a manner they are used.
2. Arrangement of individualized and unique space through seeking a relationship with nature.
3. Arrangement of individualized and unique space through seeking a relationship with tradition and culture of the location.
4. Building of spaces supporting creativity and innovative activity through raising the quality level of the surrounding area, use of quotes manifesting the act of belonging to the global culture.
5. Spaces built, the purpose of which is to integrate society.

Following global trends in architectural design, interior design, furniture design became a manifestation of acceptance of trans-national cultures. Utilizing the universal symbols and metaphors reveals a desire of belonging to the global culture, of which primary determinants are uniformized lifestyle, similar moral and social values, usage of similar symbols and metaphors. The result is a tendency to popularization of solutions in a common stylistic trend. The emerging stylistic uniformization contradicts trends of mass personalization and individualization. The coexistence of these two directions characteristic of the creative class is one of the paradoxes of contemporary living spaces.

The flexibility of living spaces can be achieved thanks to the application of kinetic architectural elements. The inclusion of mobile elements provides the possibility of reconfiguring when the architectural space that is a possibility of numerous redefinitions of functional division parameters.

Spatial changes pertaining to reconfiguration take place in a short-term or long-term time frame, while the movement of architectural elements significantly changes spatial conditions. Reconfiguration is a short-term phenomenon marking reversible changes, and the goal of which is the adaptation to the dynamically changing needs of the user (Fig. 6.2).

[2]Koncepcja twórczości *(Creativity concept)*—Edward Nęcki.

Fig. 6.1 International exposition of Salone del Mobile in Milan, Italy. Promotion of global trends, creation of "global" style in interior design and furniture design. Furniture system with personalized image on closing panel. *Photograph* A. Bonenberg

Table 6.1 A list of characteristics creating a value in information society and spatial solutions stimulating the said characteristics

Creating value in the information society	Architectural space forming goals
Ability to think creatively and devise innovative technological solutions (work space) (Fox and Kemp 2009)	– Application of formal and composition solutions enabling both the space user's intellectual stimulation as well as taking rest period breaks – Separating informal space—space for rest, isolation and concentration – Enabling to change body position and movement during work breaks – Enabling working in a group, allocation of a mutual interaction space – Enabling change of environment: exiting into outdoor space or space for relaxation – Enabling interaction with nature: quieting the mind, inspiration – Opening to the cultural landscape, possibility of interaction with others – High aesthetic quality of the architectural space affecting well-being – Applying appropriate colour code depending on the designation and role of the premises – Space arrangement facilitating work organization
Application of information and communication technologies (work space)	– Ensuring technical infrastructure enabling functioning in the information society – Where possible, inclusion of mass media messages in architectural elements: architecture as an *interface*
Individuality combined with non-standard thinking process, characteristic of the "creative class" (living space)	– Individually formed, personalized living spaces – Alteration of space layout makes it flexible to serve many designations: a creative space can be formed related to self-education, a hobby (Kronnenburg 2007)

Fig. 6.2 The author's kitchen space design solution in a residential apartment simultaneously used as an office. Poznań, Poland

Fig. 6.3 Arrangement of individualized and unique space through seeking a relationship with tradition and culture of the location

The arrangement of individualized and unique space can be obtained through seeking a relationship with tradition and culture of the location (Fig. 6.3).

The arrangement of personalized and unique space through seeking a relation with nature proves itself first of all thanks to the uniqueness of nature's art: one of the kind water basins, characteristic terrain topography, types of flora and the atmospheric climate of the location. The weather climate factors were the decisive force in the application of specific power producing solutions (water, wind, sun) as well as about the well-being of the construction materials (Fig. 6.4).

An equally important element is also space arrangement integrating society and disposition towards neighbourhood activity—in the era of distance work, "free-lancers' era", it is the quality of interaction with others and space quality that is decisive in the location chosen by people to settle in—distance from place of work is becoming a secondary factor.

In the presented these designs, the crucial element is the fact of diversified sensitivity to spatial stimuli of individual persons and therefore also on the fact that architectural space effects users with different force. In some, the impact of a too intense combination of colours will be difficulty with concentration, while others will find the same space as organized and having an interesting design.

Design of spaces for the creative class is based on supporting creative actions through appropriate creation of space. The analysis of the author's building examples verifies the need and the timeliness for the development of technologies

Fig. 6.4 Project of community in Pobiedziska, EKO-ARCH Studio Bonenberg

and confirms the fact that information revolution shapes and changes spatial behaviour. Due to the above, undertaking the subject, which combines the phenomenon of society of knowledge and the creative class as well as spatial development is substantiated. Accessibility of information and communication technologies influences the direction of spatial transformation and is characterized by:

1. Application of flexible solutions in architecture: facilitating adaptation to the dynamic changes in the manner it is used. Applying mobile elements (decomposition) and negotiable space.
2. Building space supporting creativity and innovative activities through raising the quality of the surroundings, the individualization level, openness to the cultural and nature context of architectural solutions.
3. Building spaces that integrate society within the actual and not only virtual space. Disposition on neighbour activities—in an era of distance work, it is the quality of integrating with others, and space quality is a decisive factor in where people choose to settle—distance from one's place of work is becoming secondary.
4. Positive outlook on and valuation of diversity, simultaneous multithreading in both the city culture and spatial spheres.

6.2 In Search for Identity: Advanced Technologies, Natural and Historical Values in Architecture for Creative Class

The arrangement of individualized and unique space through seeking a relationship with nature is a common strategy in newly designed public places. It substantiates itself mainly through the uniqueness of nature's works of art: one of a kind forms of natural water bodies, characteristic terrain topography, types of indigenous to the area plant life and its weather climate. Each of these elements may become the basis for a project's composition and inspiration, may decide about power production applied solutions (water, wind, sun) and may be decisive in selecting construction materials. All of the above-named factors have an impact on the final architectural shape. Combining urban development with exceptional natural characteristics increases the possibility of creating a recognizable and individualized space. A condition for success is to prominent display the characteristic natural attributes simultaneously making them available to the users. One of the many positive examples are the newly designed public spaces on the Zeeburg and Iburgh islands in Amsterdam, which will not lose their local colour thanks to the presence of water, canals and accompanying them bridges. An interesting example of an architectural concept based on using natural and cultural qualities is the headquarter offices of the Renzo Piano Building Workshop in Genoa, located in a panoramic garden. The entire object perched on a steep slope terraced according to tradition to enable planting gardens in the difficult Ligurian topographical conditions. The office building and garden repeat this unique division of terraces, in the garden creating stone escarpments, and in the building dividing the space into interior levels taken up by individual design groups. The outdoor scenic space transcends into the indoor space of the building. This effect is magnified by the glass façade. The basis of the building's design project here was greenery and the terrain's topographic properties.

An important characteristic of modern architecture and urban planning is the necessity to create design under the saturation of existing building development conditions—action in consolidated landscape with defined and significant spatial context. Dissemination of culture and customs related to knowledge and access to information as a legally guaranteed good became a fact (Thackara 2006). This directly ties to an increase in awareness of the value of culture, history and treating them accordingly is imperative, that is with respect, interest and desire to use creatively. Access to cultural resources (including historic building tissue) presents a defined value, especially to the creative class in information society. Therefore, maintaining and prominently displaying existing values is becoming an important element of modern architectural design. Actions undertaken in order to continue spatial development preserving the existing positive values preparing to revitalize take advantage of the potential of neglected areas. The arrangement of individualized and unique spaces may be manifested while seeking a relationship with the historic landscape or be based on the historic development tissue. Spaces in historic development, aside from the fact that they are carriers of local cultural values, are unique as non-standard architectural solutions. Utilization of industrial buildings being relics of the no longer existing technology and constituting a historic testimony in the given area are also a part of this strategy. Appreciation of a historic building tissue by the information society is very much justified and is included in the mass strife towards individual and one of a kind solution. Information about the past and knowledge about historic heritage engage emotions and emotionally attach the user to his surroundings. One of the indications of returning to individual architectural solutions combining value issues of nature, history and culture is the "slow architecture movement". A movement associated with contestation maintaining modern building development tempo and scale: extensive urbanization, building above measure and financial resource available. It constitutes a part of the modern "slow life" action, which is a part of the global *slow movement*. The repercussions of the economic crisis of 2008, the beginning of which was marked by the bursting of the speculation bubble on the American real estate market, forced to ask questions about the meaning of a balanced development and departing from the model of building above the realistic needs and financial resources as well regarding the issue of the obligatory residential model. The goal is to eliminate situations of over-investing and overdevelopment. This means that consumers habits as well as their up till now behaviour should be questions and brining awareness to the fact of the lack of ongoing imperative and unending increase. The *slow movement* in architecture is based on moderation and discipline while residential architecture and philosophy of life are, in this sense, closely related. In practice, this means planning construction of smaller houses than up to now and using for building them that which we already have. At the same time, good quality and durability of construction are significant. The "slow movement" in architecture consists of the following:

1. building smaller, more modest residential premises, reducing their surface area, more accurate and better definition of one's own spatial needs,
2. using, there where it is possible, natural building materials obtained in the area of the building site,
3. allowing for secondary use of elements and materials,
4. using existing spatial structures to create new ones,
5. designing in such a way as make easy the replacement/change of interior and construction elements,
6. an important aspect is using craft products and work of craftsmen,
7. placing on durability and quality,
8. adaptation to existing climate,
9. departing from the perfection of minimalism,
10. in the *slow* design process, there is time to review the project; there is time to reach for non-conventional designing methods, striving towards creating a work of art in the designing process.

A well-known proponent of this building philosophy is Eduardo Souto de Moura, a Portuguese architect whose work combines minimalism of Mies van der Rohe with a penchant for local materials and local construction techniques. In 2011, he was the winner of the Pritzker Prize—thanks to accurate, unhurried study of architecture, context and tradition. Critics also point to Peter Zumthor and his realization of Thereme Vals as an example of slow architecture. Most important, however, is that other architects are starting to follow in the same direction: architects who design on a smaller scale and it is them who can potentially have a great impact on the spatial transformation in the spirit of "slow" movement.

The counterpart of the slow approach on the scale of cities and towns can be found in the worldwide Cittaslow movement.[3] The network is aimed at sharing experiences and attempting to achieve fifty designated targets and principles—promoting a healthy lifestyle based on natural resources, contact with culture and art. It is important to raise the standard of living for all residents in the urban environment, resisting the homogenization and globalization of cities, promotion of cultural diversity and uniqueness of individual cities. The Cittaslow International Network brings together 89 cities around the world, the Polish participants of the Cittaslow network include Reszel, Biskupiec, Bisztyniec, Lidzbark Warmiński, Murowana Goślina, Lubawa and Olsztyn.[4]

References

Bańka, A. (1997). *Architektura psychologicznej przestrzeni życia—behawioralne podstawy projektowania* (p. 87). Poznań: Gemini S.C.

Bonenberg, A. (2013). *Media, przestrzeń, architektura. Transformacje przestrzeni społeczeństwa informacyjnego*. Poznań: Wydawnictwo Wydział Architektury Politechniki Poznańskiej.

Florida, R. (2010). *Narodziny klasy kreatywnej* (p. 31). [Original title: Rise of the Creative Class]. Warsaw: Narodowe Centrum Kultury.

Fox, M., & Kemp, M. (2009). *Interactive Architecture* (p. 26). New York: Princeton Architectural Press.

Kronnenburg, R. (2007). *Flexible Architecture that Responds to Change* (p. 14). London: Laurencje King Publishing.

Nęcki, Z. (1994). Percepcja środowiska—ujęcie psychologiczne. In: J. Bogdanowski, (Ed.), *O percepcji Środowiska* (p. 11). Warszawa: Zeszyty Naukowe 9, Polska Akademia Nauk, Komitet Naukowy przy Prezydium PAN, Instytut Ekologii PAN Oficyna Wydawnicza.

Thackara, J. (2006). *In the bubble—designing in a complex world*. Cambridge: MIT Press.

[3]Source: The official website of Cittaslow: http://www.cittaslow.org.

[4]Ibidem.

Chapter 7
Architectural Landscape in Mass Media Coverage

Abstract The visual potential of architecture is successfully used in media events, televised and tracked via the Internet. Creating symbols, providing backgrounds, giving the proper setting for the presented content became an important task of architecture in media message, which is necessary to create the atmosphere and drama of the performance.

Contemporary media events, from the perspective of their audience, are composed of a sequence of images. They remain the most effective medium of mass communication. Referring directly to the emotions, subconscious reactions, they have an extremely fast impact and are memorable. Spatial context, including architecture, is usually an element of media image: it can play the background role, and it can also be a carefully designed event scenery. In the first case, it shows the spatial context of events, in the second case—it shows the unreal world, but reinforcing the message of the transmitted context:

1. Architecture associated with the media event creates symbols.
2. Architecture associated with the media event creates a brand of place.
3. Architecture associated with the transmission of media helps to perpetuate aesthetic patterns introducing trends.
4. Mass media transfer affects the perception of architecture, allows to look at it from different perspectives.

7.1 Architecture as Scenography in the Mass Media Event, Supporting Creation of Place—Brand

The potential of some works of architecture is successfully used in large media events, televised and tracked via the Internet. One of the important tasks of the building may be creating the right background, the proper setting for the presented content. Architecture associated with an important event is its scenery, which is necessary to build the atmosphere and drama of the performance. Being a

A. Bonenberg, *Cityscape in the Era of Information and Communication Technologies*, The Urban Book Series,
https://doi.org/10.1007/978-3-319-69542-6_7

distinctive, easy to remember media element, it is in fact a part of the show. Spectacular architectural objects appear in the media depending on the generic structure of communication. TV transmission[1] dominated by entertainment and sport—these areas are linked to stadiums, sports arenas and halls.

During the Olympic Games in Beijing in 2008, the role of architectural icons was played by the Olympic Stadium by Herzog & de Meuron and the swimming pool Beijing National Aquatics Centre (Wattrecube) by PTW Architects. During the Olympic Games in London—this role was taken over by a swimming pool designed by Zaha Hadid.

In this context, an important issue is the reception of a work of architecture by filming and photos. The perception of spatial solutions as a two-dimensional image (even a moving image) limits the possibility of a full evaluation, including knowledge of the context of the work. Such reception can be, therefore, described as incomplete and often idealized. Construction of effects with illumination, adaptation to daytime and evening television broadcasting builds a spectacular, dynamic and perfect space. Unfortunately, reality can be disappointing, especially after the show. Images in the mass media are available in the following years. They impress other creators, thanks to which, in the distant architecture (distant geographically, functionally, conceptually) are repeated similar but not always adequate formal measures (Figs. 7.1, 7.2, 7.3 and 7.4).

Entertainment in the media appears in many different forms, including the form on everyday life and everyday human environment. In this case, it generates design trends concerning the other sphere of life—and a different spatial scale. Cultural issues, habits of many spectators are modified by the sceneries of popular television series. Using the selected colour schemes, stylistics in conjunction with the presented way of life of popular television characters has an impact on the preferences and tastes of the audience. Educational opportunities of television have been used in social campaigns for a long time. The widespread use proves the effectiveness of propagating the best practices in this way.

However, in order to sort out the problems of multi-dimensional impact of mass media communication on spatial decisions, we can relate to the theory of McQuail (2007), who defining the functions of mass media messages enumerated: the diffusion of innovation and development, social education, socialization, cultivation, entertainment and mobilization. Each area is associated with a different sphere of activity of the recipient—and this can be ascribed to the effectiveness in promoting the observed spatial patterns and motivating for change (Table 7.1).

---

[1]For example, TVP1 and TVP2. The share of auditions carrying out a mission in the terrestrial programs of TVP Source: "Report of the National Council of Radio and Television of business activities in 2010", March 2011, http://www.krrit.gov.pl.

Fig. 7.1 Olympic Stadium by Herzog & de Meuron in Beijing. Images corresponding to the medial image of the object: a wonderful panorama and arena of the facility. *Photo* A. Bonenberg

Fig. 7.2 Photos illustrating a "non-media" shot of the famous object: after turning off the lights. *Photo* A. Bonenberg

The consequences of education through communication in the mass media can relate to improving standards of architecture due to understanding the layers of meaning and symbolism, generalization of stylistic trends, the use of technical innovations in architecture, a sense of belonging to a community and culture. Contemporary architecture seems to be subjected to the above-described motivations.

7.2 New Stylistic Architectural Forms Based on the Creation and Distribution of Symbols and Metaphors

Global trends in the design of architecture, interiors, utility objects is a consequence of the existence of transnational cultures,[2] which use common symbols and metaphors. Stylistics resulting from the promoted patterns is aimed at the manifestation of belonging to a global culture. And although its existence raises disputes

[2]Transnationality is a social phenomenon and a branch of the social sciences examining the interrelationship between human beings in the context of a declining economic and social importance of borders between nation–states. The concept developed in the early twentieth century by Bourne (1916).

Fig. 7.3 Beijing National Aquatics Centre (Wattrecube) designed by PTW Architects. A striking illumination of the characteristic facade is a favourite theme of tourist photos. *Photo* A. Bonenberg

among sociologists, the mere existence of transnational cultures and ideologies is not in doubt. Their fundamental determinants are a uniformed lifestyle, similar moral values, the use of similar symbols and metaphors. It can be assumed that the transnational culture is formed mainly in metropolises all over the world that is in multinational and multi-ethnic environments. The author believes that the virtual space of the flow of information, including social media, plays a crucial role. The corporate and international nature of economic activities contributes to the application of the same spatial solutions, similar architecture and interior design in many different places in the world. The result is a tendency to popularize solutions in the common stylistic trend. The following stylistic uniformity contradicts the trends towards mass customization and personalization. The coexistence of these two

Fig. 7.4 "Non-media" view of Beijing National Aquatic

directions that are characteristic of the information society is one of the contemporary paradoxes of the living and working space.

Parallel creation of architectural forms based on the creation and distribution of symbols and metaphors is a common design strategy, especially when the recipient, consumer of architecture, is an international audience. A special feature is a reference to the collective consciousness, for which a common code of basic meanings, forms and compositions is clearly readable. Design concepts are thus based on universal message.

A specific example illustrating this issue is the architecture of the national pavilions of World Expositions (here the example of pavilions of Expo 2010 in Shanghai and 2015 in Milan), designed to build and strengthen the image of the state as a brand.

The distinctive features of the architecture of world exhibitions is its receiver—the multicultural, international community. Hence, it is crucial to refer to the collective consciousness, to which the common code for basic meanings, forms and compositions is clear. In this case, architecture is devoid of any spatial or social context. The only dimension, to which it formally refers, is cultural. For the purpose of marketing, architecture is to create a code of simple, positive and compelling associations with a specific national culture, using the universal, transnational language—comprehensible to the widest possible audience.

For example, in the Canadian pavilion (Expo 2010 in Shanghai), one can note that the building form and materials (wood, mirror steel, water) bring to mind the rough Canadian climate. At the same time, a reference is made to building trends in

Table 7.1 Socio-cultural interactions in the mass media combined with the impact on the recipient and motivations for spatial changes in the human life

Socio-cultural effects of mass media (McQuail 2007)	Effects on the audience	Architecture Themes, spatial typologies, exhibitions, strategies of national image strengthening
1. Diffusion of innovations and development. Informative function, spreading technical knowledge, encouraging individuals to change and mobility, promoting consumer ideas, education and health culture ►	Increasing the awareness of the choices made by receivers, satisfying curiosity, imparting knowledge ►	– Use of innovative solutions in architecture, – Innovative use of traditional materials, – Seeking environment-friendly technologies
2. Social learning. Correlation—explaining, interpreting the meaning of events and information, and commenting on them, supporting established norms and authorities ►	Establishing behavioural patterns, principles and good practices with regard to new phenomena. Shaping the opinion about new architectural trends ►	– Creating the aesthetic canon for modern architecture, – Transmitting the meaning of architecture, its symbolism, – Creating a modern code of meaning, semiotics
3. Socialization—teaching norms and values. Coordinating individual and communal social endeavours, achieving a social and political consensus, establishing authorities, granting social status ►	Creating a feeling of social belonging ►	– Manifesting cultural belonging through architectural form
4. Cultivation—strengthening existing opinions, consolidating conventional behaviours, the dominant culture and common values ►	Creating a feeling of cultural belonging and a community of tradition and custom ►	– Recalling cultural heritage and generating interest in the achievements of the past, – Creating link between old and new in used materials, building techniques
5. Entertainment—emotional stimulation, influencing the mood, organizing fun, leisure and reducing stress ►	Promoting lifestyle choices, including recreation and leisure ►	– Creating social spaces and free time infrastructure
6. Collective reaction—public campaigns in politics, social life, economy and other areas ►	Organizing individuals into a collective, a public good ►	– Granting access to resources: public, communal, educational spaces

the spirit of sustainable development, and the public space created in the heart of the object represents a pro-social, loyal and open image of Canada as a country.

At the same time, digitally modelled spaces with complex geometry of forms are commonly used to highlight the ingenuity and the technological sophistication of the World Exposition participants. The ability to digitally model planes or to parametrically generate the architectural forms has become a core concept for many exhibition pavilions, as the visual effects have become an indispensable element of "the architecture of the future" poetics. By relying on state of the art software, more or less influential modern architectural projects have been carried out. Architecture created from the earliest stages of development in an digital environment can occur significantly different than the structures developed in the manual tradition. In contrast to architectural hand drawing, digital modelling is an indirect form of designing: the way the software operates and its interface influence the project decisions.

In the Chinese pavilion as an example, one may observe that the form and construction materials remind of traditional wooden buildings. Nevertheless, the digitally generated geometry of its roofing utilizes good models of modern architecture.

The French pavilion uses innovative technologies for timber constructions and woodworking, exemplified in the curved wooden lattice created by XTU Architects Anouk Legendre, Nicolas Desmazières. The pavilion was constructed in a manner that allowed for it to become an exposition system at the same time. Glue-laminated timber arches were used to create a strong but lightweight structure of lattice girders and pillars. The designers used computer modelling techniques to maximize the static efficiency of all of the wooden components, which were cut using a digitally controlled robot. The elements are interlocked, minimizing the need for additional fixings.

In the United Emirates, pavilion designed by Foster and Partners, digitally modelled surfaces create an organic architectural form that resembles desert dunes. Smooth, rippling form of the building effectively combines image references to natural resources and advanced technology.

One of the most characteristic, digitally modelled architectural forms of the Milan Expo 2015 is the Vanke pavilion created for a Chinese property developer by Daniel Liebeskind. Despite its futuristic poetics, the expressive, sinuous geometry of the pavilion was not developed using parametric or generative modelling systems. The designer of the project describes it as handcrafted. The pavilion resembles the body of a dragon, covered with red, shiny ceramic scales (Bonenberg 2013).

The architecture of EXPO 2010 and 2015 exhibition pavilions is an important factor in strengthening the image of countries and nations during a global event. It influences the brand message through both spatial and architectural features, for example:

1. Use of innovative solutions in architecture or innovative use of traditional materials—for diffusion of innovations,
2. seeking environment-friendly technologies—for spreading technical knowledge,
3. creating the aesthetic canon for modern architecture—for explaining, interpreting the meaning of events and information,
4. transmitting the meaning of architecture, its symbolism—for commenting on contemporary culture,
5. creating a modern code of meaning and semiotics—for supporting established norms,
6. manifesting cultural belonging through architectural form—for teaching norms and values,
7. recalling cultural heritage and generating interest in the achievements of the past —for cultivation and strengthening the existing.

Predominantly, however, digitally modelled planes and architectural forms have become an element of "the architecture of the future" poetics, associated with the ingenuity, innovation and the technological sophistication. This more than anything supports the powerful media, image and commercial message.

To sum up, the metaphorical and symbolical message of the Expo pavilions aims to evoke simple associations, which coincide with the fundamental values of the space built for societies, the development of which is based on knowledge:

1. The reference to the protection of cultural heritage, tradition, handicrafts is understandable, among others, in forms of the Finnish (2010), Polish (2010) and Japanese (2015) pavilions.
2. The manifestation of co-creation of a modern culture through design and fashion is visible in the French pavilions (2010 and 2015).
3. The reference to the advanced technology is evident in the architecture of the German (2010 and 2015) pavilion.
4. Using forms drawn from national landscape is characteristic of the United Arab Emirates (2010 and 2015) and Qatar (2015) pavilions.
5. Placing emphasis on ecology is the conceptual base for the architecture of the Canadian (2010), German (2015) and British (2010 and 2015).
6. References to state-of-the-art digital technologies in architectural design can be found in the architecture of the German (2010 and 2015), French (2015) pavilions.

See Figs. 7.5, 7.6, 7.7, 7.8, 7.9, 7.10, 7.11, 7.12, 7.13, 7.14, 7.15, 7.16, 7.17 and 7.18.

Fig. 7.5 Expo 2015 in Milan, the Japanese pavilion. *Photo* A. Bonenberg

Fig. 7.6 Expo 2010 in Shanghai, the Norwegian pavilion by Helen & Hard. *Photo* A. Bonenberg

Fig. 7.7 Expo 2010 in Shanghai, the French pavilion. *Photo* A. Bonenberg

7.3 Place as the Subject of Fashion Created by the Media

With the release of applications on mobile devices such as smart phones or tablets via social networking, many users of social media have decided to make a note of their own presence in different geographical locations. These locations may be related to their interests, and the addressees of such information are friends in the online community. Information about where we are reaches an individual social network: it makes it easier to meet in the real world, it enables us to share our experiences, alerts about difficulties. An important aspect of the basic territorial behaviours is a choice of meeting places and entertainment—as an arena of social, group actions. At the same time, not without significance is the desire to leave a virtual "I was here" in a real and digital space.

It should be noted that being in fashionable cities, famous public spaces and surrounded by spectacular architecture becomes an element of competition and solicitation for acceptance of the community, expressed through positive comments. The opportunity to be in an appropriate spatial environment has always been an important determinant of status and a permanent element of building the image of an individual or a group of people.[3]

[3]Since the dawn of history, territorial behaviours giving rise to a hierarchy of power have been associated with the place, space. Hence the custom of making portraits on the background of the identifiable elements of the urban structure or concrete buildings. One of the first such representations were Egyptian reliefs, portraits with a suitably chosen background. Also the eighteenth and nineteenth century fashion for travelling had prestigious overtones, which was associated with an increase of image importance of education—travelling was aimed at increasing prestige in the eyes of the environment, and in the future—of the posterity. Such understanding of travelling is the basis for today's behaviours observed in social media.

Fig. 7.8 Expo 2015 in Milan, the French pavilion. *Photo* A. Bonenberg

Since the beginning of the significant increase in professional mobility, forced by the international corporate structure, can be observed a fashion for "collecting" places when travelling on business or in private. An important factor is the significant development of mass tourism. Determination on digital maps of places

Fig. 7.9 Expo 2010 in Shanghai, the Polish pavilion by design office WWAA. *Photo* A. Bonenberg

Fig. 7.10 Expo 2015 in Milan, the Polish pavilion facade. *Photo* A. Bonenberg

Fig. 7.11 Expo 2010 in Shanghai, the Canadian pavilion by Saia Barbarese Topouzanov architects. *Photo* A. Bonenberg

Fig. 7.12 Expo 2010 in Shanghai, the German pavilion by Schmidhuber + Kaindl. *Photo* A. Bonenberg

Fig. 7.13 Expo 2015 in Milan, the German pavilion by Schmidhuber + Kaindl. *Photo* A. Bonenberg

Fig. 7.14 Expo 2015 in Milan, the pavilion of the United Arab Emirates by Foster + Partners. *Photo* A. Bonenberg

Fig. 7.15 Expo 2010 in Shanghai, the pavilion of the United Arab Emirates by Foster + Partners. *Photo* A. Bonenberg

Fig. 7.16 Nemesi architects Italy pavilion—Milan Expo 2015. *Photo* A. Bonenberg

Fig. 7.17 Expo 2010 in Shanghai, the British pavilion Seed Cathedral. Designer: Thomas Heatherwick. *Photo* A. Bonenberg

which we have visited, in which he had lived, which we like, illustrating them with photos expresses an affiliation to the global culture—emphasizes a range of places that I know and I visit. The archetype of these behaviours can be found in times when travel was reserved exclusively for the people who were particularly wealthy or at the top of the professional hierarchy. "Collecting of places" via social media is also driven by visual messages coming from the general mass media: television broadcasts of important, beautiful, historical places described in the literature or shown in movies is at the core of their popularity. Mass media has been shaping the collective memory and awareness for decades. We, however, only for a few years have been experiencing the possibility of rapid recording of photos and videos, organizing data, displaying them and sharing our experiences with a wide range of people using only one small device (Figs. 7.19, 7.20 and 7.21).

Facebook regularly publishes a list of the most popular sites in the world, using the statistics associated with a function "report back". Recommendations to other users and determination on the map of important places and objects to be visited help to generate fashions and trends on individual urban districts, important works of architecture, inspiring museums. In the release of 2012, this lets us know which of the 25 global metropolises was the most popular among Facebook users over the past three years. We know from what places they most often "reported back" and thus—where they spent their time. In the category of places that are important due

Fig. 7.18 Expo 2015 in Milan, the British pavilion. *Photo* A. Bonenberg

to high architectural values, it turned out that users of Facebook, while in Paris, were more often choosing the Champs-Élysées than the Eiffel Tower, and being in New York preferred to visit the Natural History Museum and not the Museum of Modern Art. Berlin, New York, Paris, Buenos Aires, Dublin, London, Tokyo, Mexico City, Barcelona, Delhi, Seoul, Melbourne have well-known icons of architecture, buildings and symbols. Not always, however, these are the same structures that Internet users point as worthy of recommendation. In the list of "urban icons of Facebook" were sports facilities (7), public spaces (6), theme parks (2), concert halls (2) and shopping centres (2).[4] Statistics show a raising popularity of public open spaces in areas with warm climates: St. Kilda Beach in Melbourne and waterfront Wynyard Quarter in Auckland, and in cold climates become popular bars and restaurants. The survey results, however, do not contain information which the objects and places included in the above statistics carry out a permanent and

[4]Source: "Social Landmarks Arround the Word", June 19, 2012, research done from the moment of activating the function "report back" in 2010 (http://newsroom.fb.com/News/Social-Landmarks-Around-the-World-17e.aspx).

Fig. 7.19 San Francisco: an architectural icon, the Golden Gate Bridge photographed by tourists. The result of popularity is about 5,340,000 images of the bridge available on the Internet (search only in English). The bridge is used as the background in films, television series: The Rock (1996) directed by Michael Bay, the series Monk (2002–2009), the show Full House (1978–1995), which have been viewed around the world. *Photo* A. Bonenberg

Fig. 7.20 San Francisco iconic cityscape used as the background in films, television series: The Rock (1996) directed by Michael Bay, the series Monk (2002–2009), the show Full House (1978–1995), which have been viewed around the world. *Photo* A. Bonenberg

Fig. 7.21 New York: taking pictures of the city skyline. The result is approximately 790,000,000 icons associated with the New York Manhattan available on the Internet (search only in English). The views are commonly known from the iconic "Breakfast at Tiffany's" (1961) to "Sex and the City" (1998–2004). New York, *Photo* A. Bonenberg

active promotion in social media. Popularity of some of them, especially of shopping centres and restaurants is a result of active advertising campaigns.

7.4 The Perception of Architecture on Online Maps—The Fifth Facade

Using Web maps offer a new perspective of the city perception, yet this aspect is frequently underrated in designing architecture. In order to come to an unknown place without problems, it is more and more often traced in services of Google Maps, Wikimapa, Zumi or others. The place can be viewed on the satellite photo, often in a big close-up. Shapes of architectural objects, urban composition are from this perspective very clear and extremely legible. Such a view, until recently, was reserved only for observers standing in the high-placed topographic points (hills), at high architectural elements (towers, skyscrapers) or looking from the aircraft. It can be said that such a reception of the area was an exceptional situation. Today, however, thanks to new technologies, satellite images are widely used and are often the first contact of a tourist who is planning to visit the city with the specific urban space. In this context, designing the "fifth facade", a composition readable from the "bird's eye"—suddenly begins to be important. It acts as a business card of the place—it is more and more often the first thing that we see at the stage of a simple internet search, a characteristic composition, arousing associations, easy to identify and remember. A spectacular structure and proof of the potential value of such a solution is the archipelago of palms.[5] implemented in Dubai, where only a view from the top allows a full reception of the designed shape. In densely populated urban centres, what attracts attention is structures in which the geometry of the architecture is the same as the geometry of the land development.

7.5 Media Message as an Element of Architectural Space

The role of the media message, which is a part of an architectural space, is access to content, codes and cultural symbols in public spaces. This role involves two dimensions: virtual (reception of the mass media content) and the actual—identification of space due to its special characteristics: light, alternating colours, movement.

[5]Palm Islands, Dubai: Palm Jumeirah, Palm Jebel Ali, Palm Deira. An artificial archipelago of islands, landscaped mainly for commercial and residential purposes.

A parallel problem is the search for contemporary design in architecture, which increasingly involves the use of multimedia as an integral part of architecture. Many recipients and users of space positively evaluate such combinations, pointing to the values of aesthetic inspired with media technology: multimedia screens, systems of image display on the facades or illumination of objects. Impact of movement, changing colours, reflection, dynamic of changing messages increases the visibility of architectural forms—it creates a dynamic night-time panorama in the urban space. The quality of formal solutions is affected by the purpose and content of media coverage. The social functions are influenced by whether the purpose of the installation is only broadcasting of content or the interaction with the users of the space interactive elements play the role of space activation through measures aimed at entertainment, art-related, co-creating events—in which pedestrians and residents become the principal characters.

From the point of view of spatial and socio-cultural issues, public spaces are complex in nature: many interests and relationships intersect within their area, and here occurs a clash of worldviews (Weiner 2010). The purpose of the space is to establish interactions by generating experiences, feelings and events. The use of the media installation gives us the opportunity to introduce new dimensions of cooperation and deepen the existing ones. Here we can get the impression that, like social relations, they have been dominated by the mass media message, the media message sometimes dominates the cityscape. Perhaps the time will confirm the visionary thesis that the buildings will take over the role of the computer interface, and the interface will become a building (Mitchell 1996).

The multimedia message which is an element of architectural space can be divided into two main groups, due to the used media: message using a multimedia screen (of permanent nature), or light installation, projection (of temporary nature).

The use of multimedia screen as an integral part of the architecture makes it possible to extend the perception of the architectural form with the additional dimension of time. The multimedia image can highlight the nature of architecture, it can harmonize with it, taking over the role of ornament or decoration. It can also distort and disrupt it. The media installation gives to the building the basic characteristics of a living being: motion and sometimes sound (Kronhagel et al. 2006). The integration of multimedia technology and architecture represents a significant extension of possibilities of a creative expression of a designer, who includes the architecture in the space for the flow of: information, interaction, images, sounds, symbols (Castells 1999). In connection with the formation of the information society, this objective can be regarded as one of the steps towards deepening social relationships, built through the media. An example of such actions is the installation for media transmission, the Media Wall in Beijing by Simone Giostra & Partners, in which the delicate, elegant image displayed on the multimedia surface dominates the object's surroundings in the night cityscape. The similar effect has a facade of

Fig. 7.22 Touch multimedia screen on which each of the passers-by can leave their own digital graffiti—Manhattan, New York. *Photo* A. Bonenberg

the contemporary art museum by Peter Cook in the Austrian Gratz, where the light points on facades enable signalling and communication with the environment. An example of the expressive, multimedia interior was presented in the Spanish pavilion at EXPO 2010 in Shanghai.

Multimedia screens seen in the city space are an addition to the lighting system and often become the dominant element. Their use is often inspired with purely commercial activities. Due to the ubiquity and multiplicity of such elements in urban space, they are received negatively. Moving image with a commercial transmission allows for a greater impact on target groups, and it easily attracts attention (Kronhagel et al. 2006).

In contrast to screens, lighting installations and multimedia projections are of a short-lived and non-invasive nature. They are used during media events: performances, shows, demonstrations. Light is the medium and means of artistic expression. The dynamic transformation of buildings into works of art can be seen, for example, during the annual Festival of Lights in Berlin[6] (Figs. 7.22, 7.23 and 7.24).

[6]Events associated with lighting techniques have a long tradition, tending towards interactivity and mass media. One of the Europe's oldest events is the festival in Lyon, held since 1963. On 19-21.10.2012 in Łódź took place the Light Move Festival with a wide artistic program: theatre activities, concerts, multimedia installations.

Fig. 7.23 LED Screen Times Square Information Center, New York. *Photo* A. Bonenberg

Fig. 7.24 Festival of Light in Berlin: light installations whose background is the facade of the Bundestag, Berlin, 2008. *Photo* A. Bonenberg

References

Bonenberg, A. (2013). *Media, przestrzeń, architektura. Transformacje przestrzeni społeczeństwa informacyjnego*. Poznań: Wydawnictwo Wydział Architektury Politechniki Poznańskiej.

Bourne, R. (1916). Trans-national America. *The Atlantic Monthly, 118* (pp. 86–97). Boston: The Atlantic Monthly Group.

Castells, M. (1999). *The rise of network society*. Blackwell, Oxford; Malden, MA.

Kronhagel, C., et al. (2006). ag4 Media Facades, daab (p. 13). Cologne.

McQuail, D. (2007). *Teoria komunikowania masowego* [Original title: McQuail's mass communication theory] (pp. 479–491). Warsaw: Wydawnictwo Naukowe PWN SA.

Mitchell, W. (1996). *City of bits: Space, place and infobahn* (p. 48). Cambridge: MIT Press.

Weiner, H. (2010). *Media architecture as social catalyst in urban public spaces*. MediaCity: Interaction of Architecture, Media and Social Phenomena, Bauhaus (p. 95). Universität Weimar, Weimar.

Chapter 8
Place Brand-Building: Influencing Emotional Perception of Urban Spaces Through Social Media

Abstract The image of urban areas created by an appropriate urban branding can attract population, when places compete to attract inhabitants and businesses. In order to do that effectively, it is necessary to understand the potentials of place to be presented to the target groups. Implementing emotional perception category into urban assessment would break away from a certain defined scheme of urban studies, and the method can be used in creating urban branding strategies.

The diagnosis of contemporary spatial phenomena in architecture and urban planning taking under consideration changes in cultural, social and economic impact caused by the development of the information society should form the basis for creating new strategies in urban branding. De-industrialization of cities in developed countries, significant mobility of workers, development of information and communication technologies and creative industries cause changes in settlement preferences. New, *image* of urban areas created by an appropriate urban branding can attract additional population, when places compete for inhabitants and businesses. In order to do that effectively, it is necessary to understand the potentials of place to be presented to the target groups. Experience shows that not only the positive visual and spatial features attract people, but also hard to define "atmosphere", mood, city narrative, related to activities, events and history. The choice of a place/neighbourhood to live in is often based on emotions and impressions. Author proposes an urban assessment to define these features that are important, but difficult to capture. The method proposed is emotional perception. The emotional perception of urban space is a broad-spectrum experience. It can be lived through learning the city's history, myths, legends, social relations in a relation to urban space. Furthermore, the approach takes under consideration the view of specific groups of inhabitants.

In a *place brand-building*, urban emotional perception created and observed through social media can be used to evaluate:

1. predominating emotions connected to the place: curiosity, awe, boredom, distaste and irritation for place marketing and place promotion,

A. Bonenberg, *Cityscape in the Era of Information and Communication Technologies*, The Urban Book Series,
https://doi.org/10.1007/978-3-319-69542-6_8

2. city narrative: history, legend, combined with particular elements of city structure for urban branding
3. events contributing to brand building,
4. mapping emotions and presenting them on urban plan,
5. assessment how to create a powerful city brands.

The beauty of urban structures, which through the centuries was derived from the spatial order and quality of life within it, has been moved into the back row of interests of the town planners and spatial planners. Its place was taken by the values based on predomination of rational functionalism strongly related to technology and economy. Urban tradition of the first Athen's Charter introducing division into functional zones continues to be the basis for shaping modern city structures. City space was assigned with functional and economic values, with no regard to emotional values directly related to a sense of beauty or ugliness of the surrounding area. Contemporary idea of a balanced development refers mainly to the issues of nature, economy and sociology; the problematic issues of beauty are pushed to the back of the hierarchy of strategic objectives.

The reason for this are difficulties in defining and assessing beauty, both at the stage of recognition of determinants of spatial development, as well as in the phase of creating scripts for the development of the city structures. However, diagnoses, which do not encompass all significant needs of the residents (including the inbred need for beauty and harmony), imprint themselves negatively on the quality of urban planning.

It is believed that harmony of urban structures has an impact on social order and lack of appreciation for beauty is one of the reasons of urban disintegration and fading identity of cities. In the Beijing Charter of the XIX Congress of the International Union of Architects—UIA, held in June 1999, among the tasks facing the architects of the twenty-first century, there is a postulate of bringing back the soul of cities and towns which characterized them and invoked awe of its inhabitants and visitors during the past centuries.

In modern urbanization, the development of planning procedures based on functional and technological, social and economic analysis significantly outdistanced methods of diagnosis of compositional quality, order and harmony, which may become an object of admiration. The author believes that the ability to find beauty in an urban space is just as important as knowledge of the state of the environment, its physiographics, methods of land use, access to public transportation, etc. Our feelings have an objective quality in forming our relationship with our surroundings. The beauty of a city evokes specific feelings providing a sense of order.

The interesting point would be to present the possibilities of expanding existing urban diagnostic methods by adding emotional reactions as a new research tool. In urban planning, emotional perception is a rare approach to diagnostic research. It seems that it corresponds well with modern urban planning challenges but also with the necessity of creating the brand of the city in the social media.

In direct ties between a man and a city, urban interiors have basic significance. It is they that decide about perception of beauty of the streets, city squares, and housing blocks—designating contact with space. It is not hard to notice, that these components of municipal fabric play a double role. First of all, they have a specific functional designation: to meet the requirements. Secondly, their appearance evokes defined aesthetic experiences related to the form, composition, colour, etc. We can experience certain emotions while contemplating only a view of a part of the city and independently experience different feelings using its functional attributes. Thus, if we like some element of municipal space because it looks pretty, it is still not known as to whether we will be satisfied with its practical usefulness to meet defined needs.

Whilst perception of a city takes place only by means of sights, it has a contemplative characteristic; functional perception is dynamic, imposes action. It places a person under pressure of direct contact with specific people, situations, institutions, events—they provide specific experiences (e.g. looking for a parking space, taking care of a matter in a public office, finding a room at a hotel). Activity of an individual is then directed to a specific need related to part of the city's fabric. Positive or negative experiences are related to this activity, combining into a general experience of beauty of the city. The said activity evoking specific feelings can be called "experiencing" the city. Theodor Lipps (1987) says that experience is the basis for knowing and it stimulates the human psyche.

How, through this understanding, can one describe the relation between the view of a place and its designated function?

Emotional perception requires careful observation, perception and knowledge. Eyesight carries information about the external state of the observed surroundings; knowledge enables to identify with the mentality and habits of the inhabitants, their culture and history.

Emotional perception is a specific type of "listening in" to the city. The way in which it communicates to us with its appearance, history, tradition, culture, diversity of functions. It is a mental entrance into the world of the residents with their joys, sadness, aspirations and problems. It also allows to understand places marked with unusual events.

8.1 Mapping Emotions

An important factor of our emotional experiences is the feelings of relationship with the city and sensations accompanying it. They ensue not only from social relations but also from our bonds with the city space—houses, sidewalks, walls. Gradually, interpersonal elements come into play, a specific form of sympathy and antipathy. Such anthropomorphism of emotional experience, assigning human traits to houses and streets is well known to psychologists. Thus, we have cheerful and joyful streets, serious and laughing facades, sad and gloomy courtyards, and tenement houses evoking pity, pathetic, funny and anguished houses. Specific emotions ensuing from associating with a city, harmony and disharmony, city myths, fashionable and cult places, style of living related to various parts of the city—those are examples of relationships, in which the city shapes our emotional approach to its space. In a wider sense, this emotional structure of the city tissue builds a bond between a man and the space. Stanisław Ossowski (2004) points to social reaction determinants towards works of art, evoking their communication function. Through analogy, one can risk a statement that in the urban environment emotional relation towards the space also has a social function. It is based on the emotional "interlacing" of social relations, which impact the valorisation of municipal space.

8.2 Sense of Community

Emotional perception on the historical, symbolic and existential level is a vital element in building a sense of municipal community. City events, experiences, social stances, cumulative memory of the inhabitants—are an inseparable part of urban science. Perceiving and recognizing them enables obtaining urban solutions, which strengthen group ties and identity.

Emotional perception is especially useful in comprehensive grasp of a context. The key issue here is the history of a place, its identity, ambiguity. In a city space, we find various, overlapping levels of experience, which influence emotions and human behaviour. The effect of analysis is, on one hand, discovery of unrecognized contexts, and on the other hand, provision of a critical record of interdependencies, which are decisive in the uniqueness of a place.

Emotional perception is a reliable foundation for making decisions defining the scope of interference in the existing tissue of the city. Prior to undertaking project decisions a planner should ask himself the question: how the project task is going to be understood, and after formulating an answer, seek approval and confirmation within the local community.

Emotional perception monitoring may be carried out for any part of each city, allowing to draw out those values, which eludes standard urban diagnosis. Analysis conducted in this manner shows that places at first glance unattractive may take on a sparkle, unique character, its own one of a kind individuality. This is the identification of emotional urban structure topography. From this point of view, both intriguing deformity and breathtaking beauty enrich experiences constitute a place's charm and uniqueness. Analysis should be skilfully adapted to specific conditions of a location. Its various aspects may be deepened by adapting to the required accuracy of the studies. Using the approach in the field of urban studies introduces not only town planner's imagination in the design process but also interpretive diversity of emotions in urban creativity.

It is the author's belief that implementing emotional perception category into urban assessment would break away from a certain defined scheme of urban studies and the method can be used in creating urban branding strategies. Analysis of emotional perception causes urban studies to become a process emerging from the life schedules of inhabitants, requiring constant effort to discover the city and to perfect its space. Social media are the main tool of measuring, understanding and creating emotional perception of cities inhabitants.

References

Lipps, T. (1987). Psychologische studien. Weiss, Heidelberg, desc.w: James, W: Essays, comments, and reviews, Tom 15. P&F Harvard College. 391–39.

Chapter 9
Spatial Transformations and Work-Related Mobility

Abstract The increased mobility is a characteristic of the information society—migrations for economic reasons have an impact on use of space: spatial expansion, inversion and intensification are observed in the urban scale. At the same time, at the architectural level, there is a strong need for flexible design which tendencies of: reconfiguration, negotiation, adaptation.

Observing the characteristics of new urban developments, it can be seen that they are a response to trends, which are caused by economic, social and cultural change. Areas once occupied by industrial production are getting transformed into multiuse, hybrid districts with housing, office and cultural programmes. More and more often industrial activities subjected to modernization become a point of interest of the cities' urban transformation. In both cases, industrial heritage remains there exposed in preserved architectural elements or new forms inspired by industrial aesthetics—their role is to build the identity of the place.

The omnipresence of information and communication technologies not only disseminates working and living models introducing work from home modality but also it diminishes mental barriers associated with migration. The increased mobility seems characteristic of the information society as much migration for commercial reasons causing tendencies in use of space: spatial expansion, inversion and intensification are observed in the city scale. At the same time, at the architectural level, there is a strong need for flexible design: reconfiguration, negotiation, adaptation.

9.1 Expansion: Increase in Use of Space

The expansion of contemporary cities means the growth and development of urban body, and it is associated with building structures that are directly adjacent or well-connected areas: the process consists increase of the building area of the city and urbanization of nearby areas. The problems associated with the spatial

A. Bonenberg, *Cityscape in the Era of Information and Communication Technologies*, The Urban Book Series,
https://doi.org/10.1007/978-3-319-69542-6_9

expansion include functional and formal issues of new guidelines, where linking the new structures with the existing urban assumptions is of equal importance. An important theme here is natural, environmental and economic relationships: sustainable urban management, sustainable transport, construction and urban design. The objective of the European policy in this regard is to create high-quality urban areas through urban environment management in terms of social, economic and environmental issues (Kozłowski 2006).

Unfortunately, one of the forms of the spatial expansion of cities in many places around the world (including Poland) is still the urban sprawl, the phenomenon of a rapid, poorly controlled growth of buildings with a low intensity over large areas. This raises the number of known economic, social and spatial problems. But even with ordered planning of higher-intensity developments, the lack of spatial identity is often a problem. The absence of historical, social and spatial references and poor formal quality of architecture make the definition of urban centre very difficult. One example illustrating the lack of this relationship may be the continuation of construction on the waterfront in Miami Beach (USA) (Fig. 9.1).

Fig. 9.1 Ocean Drive frontage with the dominant buildings in the *Art Deco* style. *Photo* A. Bonenberg

Fig. 9.2 Expansion of building structures in the northern section of Ocean Drive in Miami Beach, the continuation of coast development without maintaining a historical scale of buildings. Author's drawing analysis emphasizes the monumental character of the new architecture built along the shoreline

Here, the primary axis defined by the Ocean Drive embankment, dominated by residential villas and hotels built in the graceful style of American Art Deco,[1] was closed by the cluster of seventeen-storey apartment buildings and hotels at a very large scale, built on guarded, enclosed areas. Even though the solution designed and completed in 1999 by Michael Graves is described by its creator as alluding to the context of Miami Beach by "spirit" and "colour"[2]—in fact, it is obviously contradictory to it (Fig. 9.2).

A positive example of the development of the waterfront is the expansion of housing developments on one of the islands of the Venetian Lagoon, Mazzorbo,

[1]The development of Miami Beach city as a known spa took place in the thirties and forties of the twentieth century. Structures in the Art Deco style are the dominant part of the historical buildings of the city.

[2]Based on the website of the architect: www.michaelgraves.com.

Fig. 9.3 Development of the island Mazzorbo by Giancarlo De Carlo (1980–1997), Venice, Italy. *Photo* A. Bonenberg

being a continuation of the well-known tourist destination: the island of Burano. The design by Giancarlo De Carlo was created in the period from 1980 to 1997. Similarly as in the case of Miami Beach, the buildings have primarily a residential function, complemented by the services, but the scale of the project is much smaller —there were constructed only 36 apartments. Here were created public spaces with communal green sites. The project has been developed with a great attention to landscape. Ulike in Miami developments, the scale, forms and colours of buildings maintain and support a unique, local character. However, the question of the context in the project is not over with that: an important research topic of Giancarlo De Carlo was studies on the insular tradition of living and adaptation of the spatial layout to the habits of future users. The issue of context was, therefore, extended with a cultural and social aspect that makes this project unique (Guccione and Vittorini 2005) (Fig. 9.3).

9.2 Inversion: The Loss of Function

A shrinking city can be seen in the demographic aspect (population outflow from the city), economic aspect (deindustrialization, unemployment) and spatial aspect (degradation of the residential or industrial tissue, leaving the site by users). The primary causes are dynamic economic changes, and the phenomenon that occurs on a large scale is shrinking of industrial areas for the development of other functions of the city. The tendency to move the industry out of European cities in the 80s and 90s was associated, among other things, with the transfer of production in the geographical zones of a greater cost-effectiveness. Searching the concept of redevelopment of wasteland and changes in the profile of use are the current solution. Industrial areas are converted to new districts of creative services with cultural, educational and residential facilities.

A competitive solution can be a renaturalization of areas. The restoration of the natural state of the environment, as close to the original state as possible, will increase the ecological quality of the area.

9.3 Intensification of Development and Transformations Within the Existing Structure of the City

The third type of transformation of urban spaces, or intensification, involves replacement of old buildings with new buildings, considerably increasing the intensity and height of building structures. In the process of intensification can be used plots that have not been invested so far, which in urban areas is rare. The intensification is caused by increase in land prices in global urban agglomerations and the image and political motivation. The effects are especially evident in countries experiencing the economic boom: China, India, South American cities.

The intensification of building development on a large scale is criticized in China because of the demolition of *siheyuan*—the traditional typical buildings and due to the massive displacement of thousands of people living in them. Lack of planning and spatial policy, practice of quartering population to others, lasting from the 50s led to a dramatically low standard of living in *siheyuan* in Beijing (Greco and Santoro 2008) (Fig. 9.4).

Spatial consequences of intensification of buildings in China are analogous to the urbanization of cities in the USA and Australia, which took place a few decades earlier. The dynamic panorama of the New World is the result of these tendencies and ambitions. The symbolic dimension has here a global race in terms of the altitude, associated rather with political aspirations and a specific image than the actual needs (Fig. 9.5).

In cities with the large-scale intensification of building development, the perception of space is specific. What strikes here is a particularly strong diversity of impressions and reception of the city from street level and from high altitude: the

Fig. 9.4 Intensification of the downtown development in Shanghai: the zone of towers forming a panorama of modern, dynamically developing city; in the foreground, the street zone, with a perceptible atmosphere of the old town. *Photo* A. Bonenberg

Fig. 9.5 Tallest building in Shanghai, and the fourth highest in the world: Shanghai World Financial Centre (2008). *Photo* A. Bonenberg

Fig. 9.6 San Francisco: The panorama of the city resulting from the intensification of development. View from the Pier. *Drawn* A. Bonenberg

zone with the first few floors of buildings is the one including fragments of the old city. Here goes the life of the street, and here we commune with the spirit of the place. The second zone is a new space—associated with high buildings, sky and air. In the major cities like New York or Shanghai, the great impression is made by the change of the place character, depending on the perspective from which we see the city (Figs. 9.6 and 9.7).

9.4 Adaptation of Existing Structures

Within the rapidly changing cities, existing building tissue undergoes transformations as well. The basic type of which is adaptation or reuse of existing buildings or spaces for a new purpose. In terms of historical background, architecture and interior design were adapted both to new function, but also to follow the new fashions and stylistic trends. It resulted from the evolution of design, aesthetics and technical capabilities. The main factors for the change of use are the evolving cultural context and economic situation.

The directions of changes proposed by contemporary creators when adapting a historic architecture can be diversified, so it seems necessary to recall important examples of adaptation in the European architecture. The culture that fromed the strict hierarchic social and moral rules of the nineteenth century have come a long evolution towards openness and transparency. The essence and metaphor of this process in archtectural dimension can be reperesented by the reconstruction of the German Reichstag. Glazed, lightweight design by Norman Foster is available for

Fig. 9.7 San Francisco: The panorama of the city resulting from the intensification of development. *Drawn* A. Bonenberg

people from the outside—citizens and allows visual contact with the interior of the object. This form stands in stark contrast to the visually heavy, neo-Renaissance walls of the building bodies.[3] In combination with each other—both of these qualities are testimony of the change in the perception of the role of "politically engaged" architecture, the more that the project also applies to other important contemporary issues: environment protection and sustainable development. In the facility was used a number of pro-ecological solutions: the design of the dome gives a natural light to the plenary hall, and it has an integrated heat recovery system and supports the retention of rainwater. In this sense, it determines the paths in the development of architecture, which can be partly used in solving the problems of the modern world (Fig. 9.8).

Equally profound changes affected the trends of shaping the place of residence—contemporary adaptations of objects can refer to the directions of reconfiguration and negotiation and are often associated with the reconstruction of objects to allow multiple and variable use of space.

Adaptation associated with economic changes aims at requalification of the former industrial areas for residential purposes, services and cultural facilities. This trend is evident in many countries, and one of the most interesting examples is the

[3]*Plenarbereich Reichstag*, the seat of the German Bundestag, originally built in the neo-Renaissance style based on the project by the architect Paul Wallot in the years 1884–1894. In 1933, this building was almost completely destroyed in the fire. In the 1960s, it was rebuilt, and in the years 1991–1999, it was readapted by Norman Foster.

Fig. 9.8 Reichstag: juxtaposition of traditional building tissue with the modern dome by Norman Foster. *Drawn* A. Bonenberg

District 798 Art Zone in Beijing, which is a post-industrial area converted into an arts centre (Greco and Santoro 2008). A large part of the buildings in the area were established during the China's first five-year plan. Since 1951, there had been created here a giant armaments conglomerate, built in collaboration with experts from the Soviet Union and the German Democratic Republic in the framework of the unification of the socialist bloc. The architectural part was designed by the Germans, who proposed the development consistent with the spirit of the Bauhaus. Quotations of the chairman Mao are visible in the interiors to this day.

A distinctive, exceptional formal and functional quality of this space was obtained thanks to:

1. Technical infrastructure of factories running out of buildings and along the main thoroughfares of the plant. Tubes, pipes, elements of machines were not disassembled after the reclassification of the land.
2. There has been preserved the building structure of varying altitudes and building materials, supporting the impression of the spatial disorder. Multidirectionality of actions of users of this space (galleries, shops, cafes, manufactories) deepens the complexity of the spatial situation.
3. Free and variable composition gives a sense of liberty and creative freedom (Figs. 9.9, 9.10, 9.11 and 9.12).

Fig. 9.9 Pedestrian passages in the District 798 Art Zone in Beijing. *Drawn* A. Bonenberg

Fig. 9.10 Small-scale art shops in the District 798 Art Zone in Beijing. *Drawn* A. Bonenberg

Fig. 9.11 District 798, Beijing, China. Adaptation of the armaments factory area and post-industrial buildings located on the premises into art galleries, commercial facilities, small gastronomy. *Photo* A. Bonenberg

Other examples in a smaller scale include the "Tramway" in Glasgow, Scotland —the adaptation of the tram depot into the cultural centre and "Manufacture" in Łódź, the adaptation of post-industrial facilities into commercial.

9.5 Reconfiguration—Kinetic Architecture

Reconfiguration of the architectural space means the possibility of multiple redefinitions of parameters for the functional division, based on the mobile architectural element or the elements of interior design. Changes associated with the spatial reconfiguration take place in the short-term or medium-term time frame (Fox 2009). Reconfiguration of the architectural elements can be observed both at the scale of a building and of an architectural interior, and Renzo Piano made use of these options in the development of the office and exhibition spaces in the project of museum in Trento and in the revitalization plan of Porto Antico in Genoa.

Fig. 9.12 "Tramway", Glasgow, Scotland, an example of a tram depot adapted into the cultural centre. *Photo* A. Bonenberg

The use of multifunctional interior architecture was not uncommon. One of the first and basic examples of creating a kinetic, reconfigurational house interior is a historic villa Schröder by Geritt Rietveld (1924). In the layout of the first floor, the following are defined: the stairwell, kitchen and bathroom, and the remaining spaces are delimited depending on the needs of users. This classic example, created in a cultural reality completely different from the contemporary reality, was followed by the contemporary projects.

The exterior elements of reconfiguration can manifest the dynamic nature of the internal spaces of the building, if there are applied moving, mobile facade systems, sun and noise protection systems. The exterior of the building should manifest its internal order. According to the author, to a group of kinetic architecture, we can qualify these objects, where the movement of architectural elements significantly changes the spatial conditions. Reconfiguration is a short-term phenomenon, meaning reversible changes, designed to fit to the rapidly changing needs of the user. Adjustment may be based on the multiplication of modular components, and it can also result from changes in the geometry or location of elements that make up a particular object.

Thus, the change is visible both in the usage and perception of architecture. An example of architecture employing reconfiguration as an impacting factor can be Starlight Theatre by Uni-systems (realization in 2003). Motion affects equally the interior and exterior of architecture—its nature, aesthetics and philosophy of creation and use of the facility.

In the other way works the multi-purpose visual arts centre Stewarts Melville College in Edinburgh, and the author collaborated in elaborating this project. Reconstruction of the nineteenth-century school assembly hall was intended to improve and increase the possibility of using the space. Flexibility of the space was achieved through the system of movable auditorium modules. This offered four variants of use: examination (open space: without the stands), spectacles (with all the stands open) and indirect variants. An example of the centre suggests that the interior can made responsive primarily due to elements of equipment. The stands were available on the market as one of the standard solutions. Functional objects must be designed to enable their transport, so they should be light and mobile (Fig. 9.13; Diagram 9.1).

In the field of interior design, the possibilities of reconfiguration are often determined by elements of interior furnishing. Indeed, the role of certain items of daily use is to optimize the architectural space associated with the variable nature of a human activity. It helps to meet the conditions of ergonomics and stress the aesthetics of the interior in conjunction with the cultural, economic, environmental and technological phenomena. In these areas, there is a special opportunity for innovative solutions in the issues oscillating around architecture and design.

The interest and the role of the designer should go far beyond the issue of form. The following conceptual designs have been created under the supervision of the author, attempting to aply theese concepts. In the presented works, the most important objectives and characteristics were requirements for flexible use of space through the use of moving parts of structure, but also topics of ergonomics, safety of use, possibility of transport, storage, packaging, which in the case of mobile elements are more complex (Figs. 9.14, 9.15, 9.16, 9.17 and 9.18).

Fig. 9.13 Stewart's Melville College—a responsive interior architecture—a cooperation on the project by Simpson & Brown Architects Edinburgh, Scotland

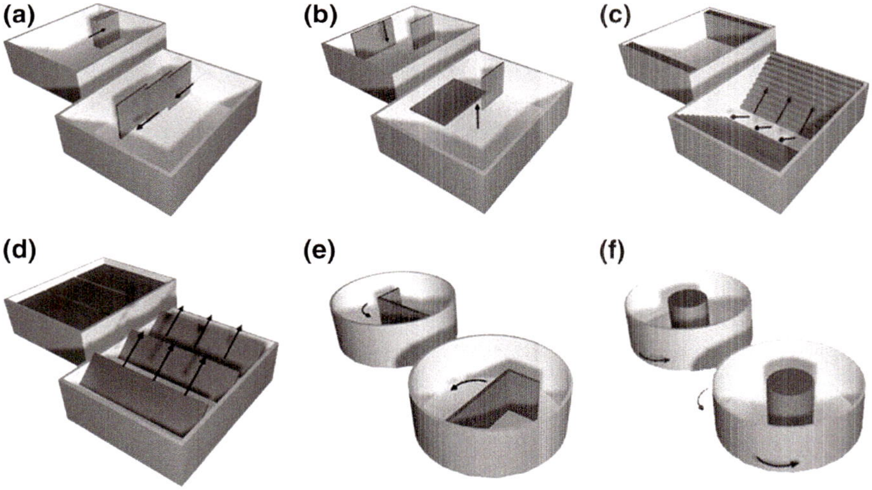

Diagram 9.1 Different types of kinetic elements in the interior architecture: **a** sliding walls (residential and office functions), **b** folding mezzanine (residential functions), **c** folding audience system (rooms for performances, sports), **d** rotary system (scenic restaurants), **e** folding roof (concert halls, sports facilities). Created by A. Bonenberg

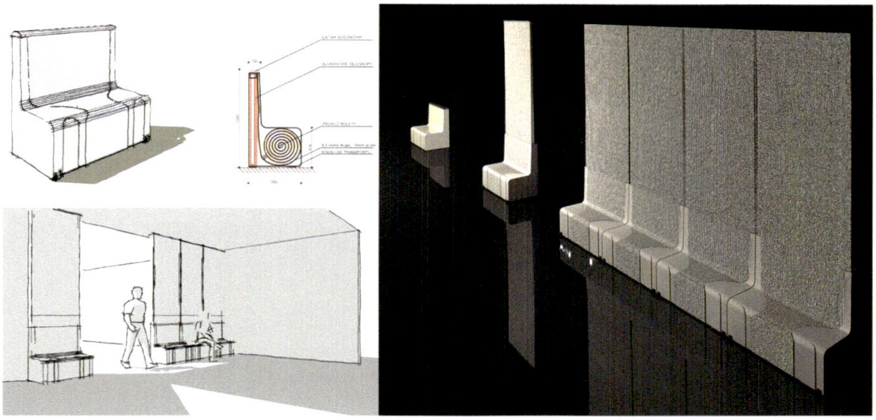

Fig. 9.14 Modular movable wall system. *Design* Michał Marchewicz under supervision of A. Bonenberg at Poznan University of Technology

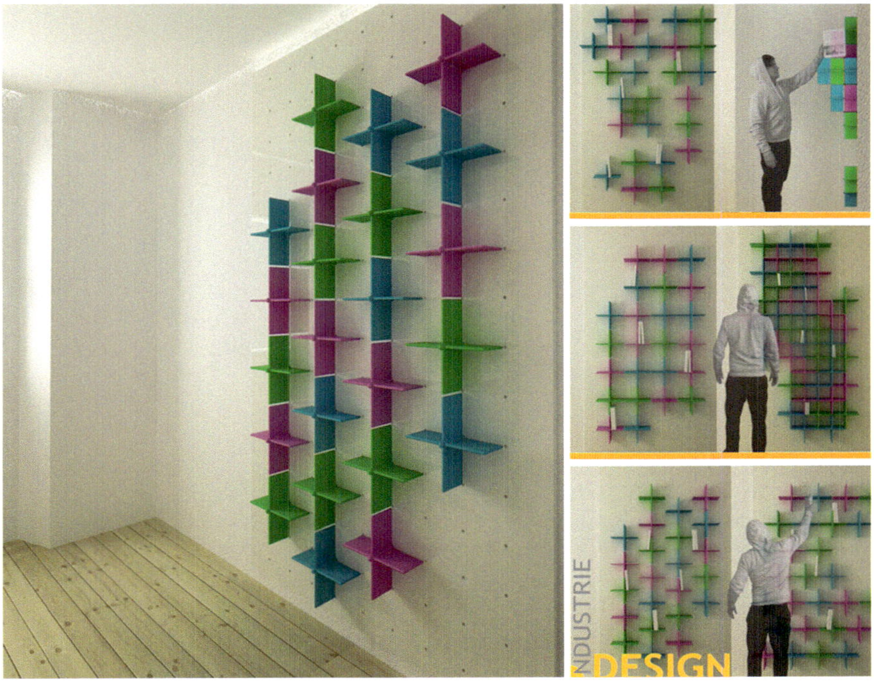

Fig. 9.15 An example illustrating the customization options of the wall shelves system, whose layout can be significantly personalized. The solution involves a diversified loading of system during use. It provides an easy access to articles of different sizes. The project executed under the guidance of the author by a student Giovanni D'Angelo within the subject of Design at the Faculty of Architecture at Poznań University of Technology

Fig. 9.16 Conceptual project of furniture for multifunctional space: kitchen system. *Design* Wiktoria Fryca under supervision of A. Bonenberg at Poznan University of Technology

9.6 Negotiation—Changes of a Functional Space Division

Negotiation in architectural space means the ability to redefine the functional separation, which takes place in a short period of time. Spaces for negotiation are created around the static compositional systems, designed so that the space could be used in many ways. In terms of discussion about the characteristics of works of

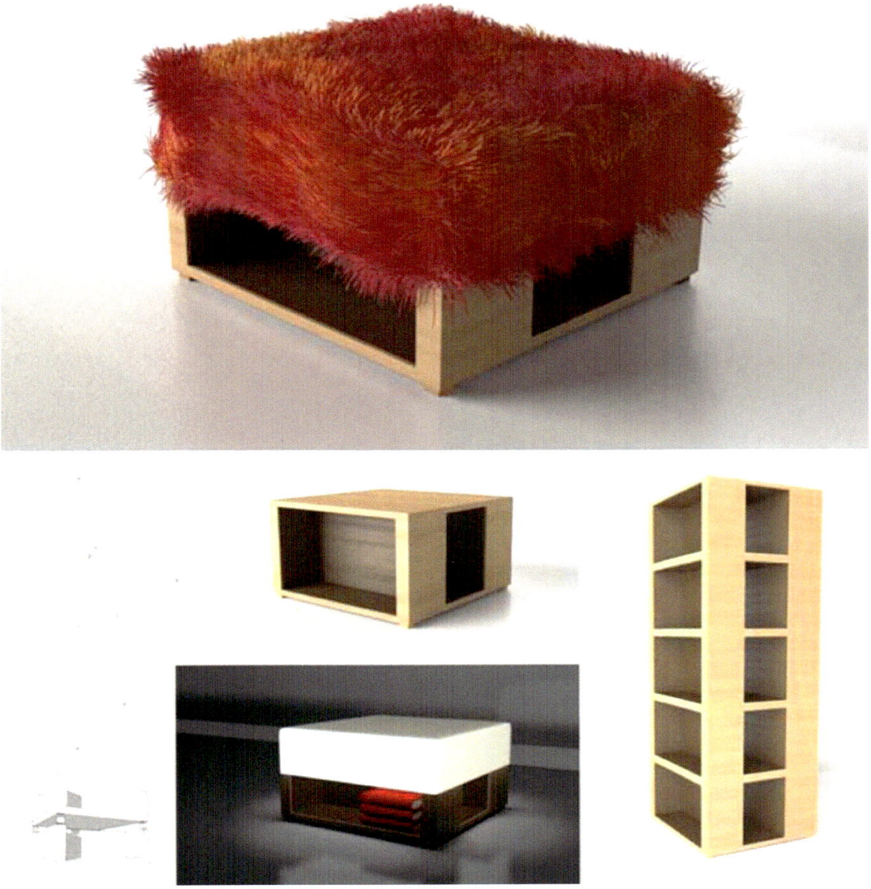

Fig. 9.17 Conceptual project of furniture for multifunctional space: Change in! modular system. *Design* Małgorzata Wawrzyniak under supervision of A. Bonenberg at Poznan University of Technology

architecture, we can talk about the negotiation of space with regard to the situation in which the usable space is not assigned to a specific function, but it offers the opportunity to use it as associated to one purpose, then to another one. Negotiation in architectural space means The ability to change the proportion of space occupied by a particular function for a short or moderately short period of time.

Fig. 9.18 Conceptual project of furniture for multifunctional space: Kulto—mobile mattress with a night table. *Design* Mirosław Wojcieszak under supervision of A. Bonenberg at Poznan University of Technology

In contrast to the recomposition, it does not consist in generating changes in a given space. In practice, designing the space for negotiations is connected with the arrangement of an open plan of areas around certain fixed functions such as stair risers, bathrooms and kitchen.

Classic examples of this approach can be Farnsworth House by Mies van der Rohe from 1950 and Glass House by Philip Johnson from 1947. These historical examples led to the popularization of the "negotiating" space in the newly designed buildings, especially that the concepts of the spatial negotiations converge in many points with the slogans of minimalism (Figs. 9.19 and 9.20).

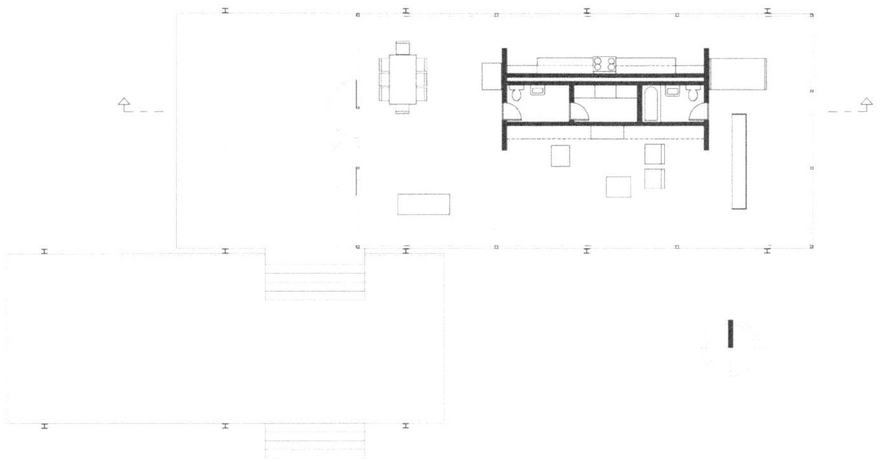

Fig. 9.19 Farnsworth House: the body and the ground floor layout

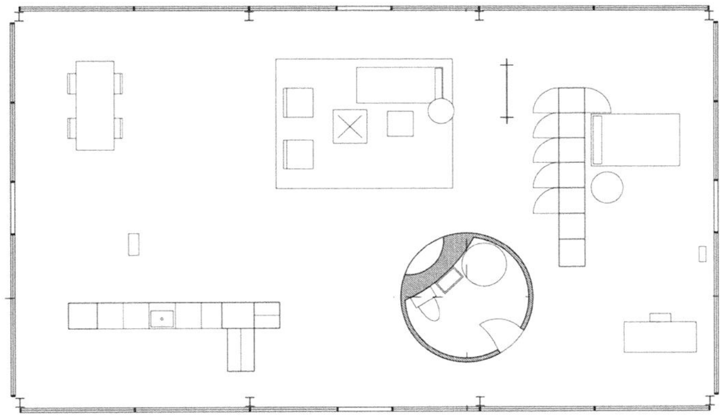

Fig. 9.20 Glass House Philip Johnson: the body and the ground floor layout

Spatial composition of residential buildings is adapted to the cultural norms being the consequence of lifestyles and conditions of employment. The transition from layouts of separate, defined areas to flexible spatial arrangement has become a reality, the same as a gradual departure from the model of the multigeneration family. A growing number of people with one child only, or living alone influences the typologies of housing; the avarage size and layout are undergoing transformations. There is a social pressure to build "perfect" mutual family relationships—hence the spatial layout is open, focused. It fosters close, intimate relationships.

As in the case of the recomposition spaces, also in the case of the spaces for negotiation, an important issue is to design elements of the interior around which the negotiation will be made. They can be architectural or interior design elements. The shape of the solid element affects the type of relationship between the users and the space.

Elements may be different from each other:

1. By form—oval and circular are conducive to increased joining of functions. In contrast, rectangular elements are divided in a more definitive way.
2. By altitude—full-height forms delimit the space better than low forms; interesting effects can be achieved with complex forms, with slots openings or cracks.
3. By transparency—the divisions may be made wholly or partly of openwork, translucent, transparent or completely transparent elements (Diagram 9.2).

There are examples of projects that combine the features of reconfiguration and negotiations. One of them is the conceptual design of the object of a temporary residence on the island Iburgh in Amsterdam by A. Bonenberg and his team: W. Janik and M. Krasodomski. The basic problem was connected with the need for privacy in open spaces. Hence, the work was preceded by research and classification of activities that the average user will be happy to perform in the company of other people, and for which he needs the maximum privacy. As a result of the study were developed three basic groups of space:

1. 100% flexible—for the public activities, spaces fully negotiated, spaces completely open, destined for commercial activities,
2. 50% flexible—for informal socializing, family, work performed at home—this is a reconfiguration space—partly closed with a sliding wall.
3. 0% flexible (essentially inflexible)—traditional spaces divided into rest areas, family life areas, personal areas.

The division was postponed at the architecture of the building, which arose on the basis of the interpretation of the results of the mentioned study; three floors, ordered from the most open to the most closed system (Figs. 9.21 and 9.22).

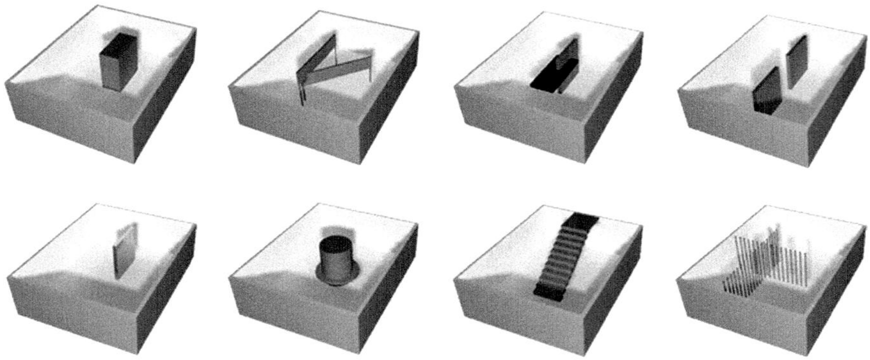

Diagram 9.2 Negotiated spaces: interior dividers. Created by A. Bonenberg

Fig. 9.21 Temporal visualization of the project Iburgh temporal living. A. Bonenberg and his team: W. Janik, M. Krasodomski

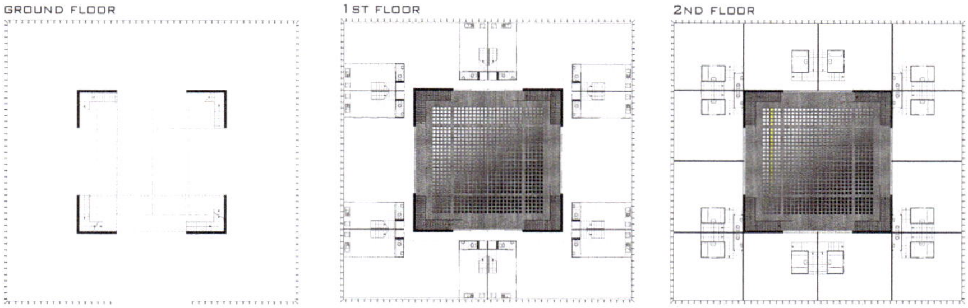

Fig. 9.22 Layouts of three storeys of the project IJburg temporal living: negotiable, recompositional and traditional by A. Bonenberg with his team: W. Janik and M. Krasodomski (Bonenberg, 2013)

It is possible to trace the spatial directions and transformations of the society in which relationships are formed on the basis of access to media information. In this regard, six notions concerning spatial changes, reconfiguration, negotiation, adaptation, expansion, inversion and intensification, replace the Vitruvian triad (Vitruvian virtues): the unity of function, structure and form. In historical regard, the purpose and goal of a building was yet durability allowing protection against the hostile natural and human forces. Immutability was the biggest asset of building structures built in order to survive the centuries. This idea was expressed by the formal solutions of the building, highlighted by the characteristics of an ornament. Sponsors, investors of palaces and temples belonged to a small group of the privileged ruling class, for which was important tradition, continuation, continuity and transmission of heritage. Examples of this approach are commonly seen around —the composition of palaces of the Florentine families even today gives us the impression of stability and balance through symmetry, horizontal articulation of rustication, size and proportions of window openings. In the past social conditions, the dynamics of changes in customs, and hence the specialization and the need for new types of space evolved more slowly than today. In the theory of architecture was established the Vitruvian feature of durability of an architectural work as the intrinsic characteristics of a great architecture.

Early examples of the flexible, variable architecture are usually associated with either a nomadic lifestyle (structures of yurts, Bedouin tents or Indian tipi) or come from a non-Western cultural background. Indeed, one of the most beautiful villas in Japan is Shugakuin Rikyu built in 1659 for the Emperor Gomizuno. The villa is an ideal of simplicity, lightness, flexibility and the integration of the internal and external spaces (Kronnenburg 2007). The use of the same premises in many ways was also popular in Europe; it resulted, however, from economic grounds associated with a deficiency of space. The model with a far-reaching specialization of individual rooms was a consequence of fashion introduced by the residences of rulers and wealthy merchants (in urban conditions). The revolution in the approach to constancy in the evaluation of an architectural work took place in the era of modernism, although the structures of free layouts and free facades conditioning the creation of flexible structures can be found already in the Art Deco buildings.

The changes described by Alvin Toffler in "Third Wave": a situation of a continuous technological evolution, changes in attitudes, trends, the need for innovation has a significant impact on the architecture. The relationship between the space to live, work, services and users is particularly dynamic. Consequently, we can observe two basic trends. The first is the construction of objects with a pre-defined short period of use. The construction costs are minimized, and after the specified technical death of the building, it will be replaced with a new one—meeting requirements and corresponding to aspirations of the new owners. Radicalizing the picture of such a situation, Rem Koolhaas says that following this trend, we can become the first generation of creators who will not leave spatial traces in the perspective of several centuries—the architecture older than 50 years will undergo the programmed technical death.

The second strategy is to build multifunctional architectural spaces, with the assumption that the characteristics of flexibility and responsiveness may protect against the devaluation of the object in the conditions of continuous changes in the environment of the recipients of architecture. This second alternative satisfies the need of leaving a material "souvenirs", although the desire to create a flexible architecture requires more economic investment. The strategies are applied depending on the role to be performed by a building. Today we accept the variable nature of works of architecture and modifications that may occur in it—already at the design stage, we anticipate potential future uses of the designed space. We conduct experiments on flexibility of its functions: ability to reconfigure the space, its negotiability and possibilities of its adaptation. Never before was the possibility of variability and flexibility such a desirable attribute of space shaped by men.

In the work "Interactive Architecture", the need to deepen the study of spatial flexibility in buildings is explained by the creation of the information society: "*there was a time, not so long ago in evolutionary terms, when our lives were based on the possibility of movement and flexibility, in fact, we owe the survival of species to that skill. Nowadays, the majority of modern cultures runs more or less sedentary lifestyle, but it may result that flexibility will once again become a priority in the development of man and technology, the social and economic changes enforce, or at least support a new form of nomadic existence: based on world markets, the global Internet network and a cheap and fast transport*" (Kronnenburg 2007).

The need to analyse topics associated with the changing destination of buildings is in a bright and humorous way raised by the message of the arrangement of the Polish pavilion awarded the Golden Lion during the Venice Biennale in 2008. Landmarks of the Polish latest architecture thanks to the photograph collage by Kobas Laksa receive brand new features, far from the original ones, which affects their external appearance, the aesthetic level and quality of the immediate environment. New features reflect economic and social trends and those relating to the spatial development, which can show up in the future. And thus, the office building by Norman Foster in Warsaw is amended into jail, which would be the effect of the fall of the city centres, and the new terminal of the Okęcie airport is changed into a farm of animals, which was justified by the authors with the forthcoming energy crisis and the collapse of air transport while simultaneously having to produce food in a short distance from the place of its use. These examples illustrate the issue in an exaggerated way, but they touch the heart of the relationship between architecture and its feature that often turns out to be a very variable and dynamic. Spatial transformations can be divided into qualitative and quantitative, and these concerning the quality are associated with architecture, and quantitative changes—to the urban development (Diagram 9.3).

The changes that without altering the size of the developed area result in the optimization of use of space may occur at a certain time, so we can distinguish: long-term changes (the timescale of several years or a few decades of years), moderately long-term changes (the timescale of a few months) or short-term changes (the change can occur in one day). The timescale is therefore one of the

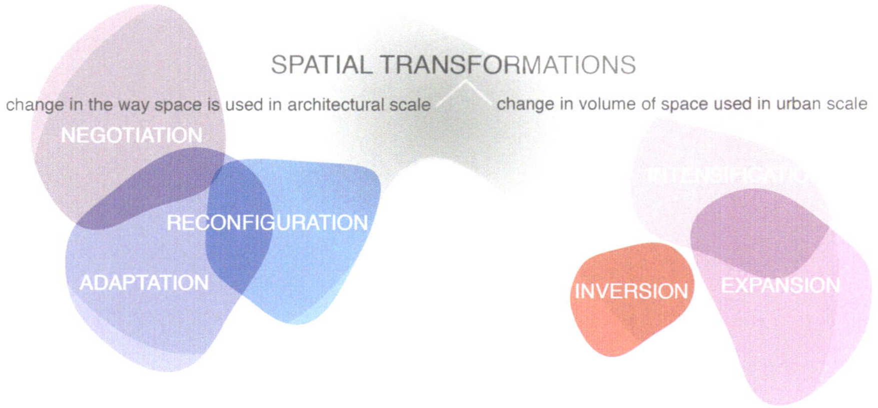

Diagram 9.3 Diagram illustrating the scheme of spatial transformations by A. Bonenberg

parameters that help to classify the phenomena of changes in the quality. Because of it and the manner of use of the space, the initial distribution of changes in quality and composition of space can be represented as reconfiguration, negotiation and adaptation.

References

Bonenberg, A. (2013) Media, przestrzeń, architektura. *Transformacje przestrzeni społeczeństwa informacyjnego*. Poznań, Wydawnictwo Wydział Architektury Politechniki Poznańskiej.

Fox, M., & Kemp, M. (2009). *Interactive Architecture* (p. 26). New York: Princeton Architectural Press.

Guccione, M., & Vittorini, A. (2005). *Giancarlo De Carlo*. Electa, Milano: Le ragioni dell'architettura.

Greco, C., & Santoro, C. (2008). *Beijing the new city* (p. 147). Milan: Skira.

Kozłowski, S. (2006). *Żywiołowe rozprzestrzenianie się miast, narastający problem aglomeracji miejskich w Polsce*. Białystok, Lublin, Warszawa: Wydawnictwo Ekonomia i środowisko

Kronnenburg, R. (2007). *Flexible architecture that responds to change* (p. 14). London: Laurencje King Publishing.

Chapter 10
New Urban Developments in Genoa, Milan and Trento as an Illustration of Spatial Transformations Designed in Response to Change in Socio-economic System

Abstract The common feature of presented case studies is the innovative approach to the problem of the urban continuation and relationship between the old and new urban structure. Three recent urban projects designed by the architect Renzo Piano refer to the revitalization of former industrial areas in three Italian cities: Genoa, Trento and Milan (Sesto San Giovanni). The quality expressed in emphasizing the unique, individual characteristics is associated here with the industrial and historical heritage, natural environment and contemporary design.

New trends in the development of the space of the society of knowledge have been visible in projects and realizations for sometime. It is important to isolate the characteristics of selected research examples, according to the methodology of research by design and to define the dominant directions of spatial transformation.

Three recent urban projects designed by the architect Renzo Piano refer to the revitalization of former industrial areas in three Italian cities: Genoa, Trento and Milan (Sesto San Giovanni) All projects constitute the material for observation of the phenomenon of transformation of urban spaces and architecture design of post-industrial era—which means entering into relationships that are characteristic of the information society.

Project tasks are a tangible result of stopping or reducing employment in heavy industry[1] for the development of alternative economic sectors, mainly services, new technologies, trade. A direct consequence of these changes is the implementation of the project of a new living quarter in Trento and the Sesto San Giovanni design in Milan—in both cases, it is aimed at requalification of post-industrial areas. In contrast, the solutions adopted for the development of the city of Genoa assume the coexistence of industrial production and allow development in the spirit of the second reindustrialisation, which is increasingly discussed in Europe, immersed in crisis.[2]

---

[1]Sesto San Giovanni is a territory of the former steel works, the district of Trento is an area formerly occupied by the factory of tires Ex-Michelin, while the Genoese Waterfront is dominated by the waterfront shipyards and ports.

[2]The necessity of reintroducing the industry to France and Europe was the main issues of the elective speech provided by the French President François Hollande (2012).

A. Bonenberg, *Cityscape in the Era of Information and Communication Technologies*, The Urban Book Series,
https://doi.org/10.1007/978-3-319-69542-6_10

All three concepts are inherent to the creation of urban spaces for the information society—the conscious society in which the selection of the place of residence is made on the basis of the space quality. The quality expressed in emphasizing the unique, individual characteristics associated with the location and the natural environment, cultural heritage in the form of reminiscences of the industrial era, the broader environment-friendly concept of creating and use of sites. The growing importance of local public spaces, strengthening of neighbourly ties enforces the need to integrate recreation areas in the urban development projects, to create high-quality public neighbourhood spaces. The common feature of these projects is the innovative approach to the problem of the urban continuation, neighbourhood and interpenetration of the old and new urban structure. The observed spatial transformations are often associated with the desire for flexibility and easy adaptation of sites to different methods of use. The problem of transformation of the urban space of the information society will be presented using the examples of works by Renzo Piano, due to the author's participation in the work on these projects and authorial contribution to the solutions used in the Trento (Figs. 10.1, 10.2 and 10.3) and Genoa Waterfront project.

Developed areas of Genoa and Sesto San Giovanni constitute an important spatial and social issue.[3] Depreciated buildings and used technical infrastructure are an economic burden for municipalities. Attractive location of these areas is a potential source of qualitative changes, and it gives the chance to create the image and regain space for residents. In each of the above-mentioned cities, exists the opportunity to replan a high-quality space of social communication, an integrating public space: recreational green areas, educational facilities and commercial buildings. In the approach of Renzo Piano, the spectrum of the created project proposals is the result of the expectations of customers and individualized treatment of space, with all the wealth of cultural, natural and social resources, taking into account economic conditions. Each of the projects emphasizes the most characteristic features of the place, creating an attractive media image: easy to identify and remember. In the proposal for Sesto San Giovanni appears the motive of industrial tradition of foundries in Milan, in the architecture of Genoa one can feel the inspiration by modern, port character of the city, in Trento architecture is inspired by the natural landscape. Despite many differences, post-industrial spaces in Genoa, Trento and Sesto San Giovanni were characterized by a number of common features. Negative features include:

1. Partial or substantial spatial separation of post-industrial areas from the city by railroads which provided the raw materials for factories and by the water necessary for the processes of production and transportation.
2. In all three cases, access to the factories was operated by the railways; railway infrastructure is still there, and in the case of Genoa, it functions as an active regional communications line. The belts of road and railway communication separate new areas of design from existing urban structures. Linking the new

[3]The project in Trento is already in the final phase of implementation.

housing with the existing urban organism is therefore difficult, especially in Genoa and Trento.

3. The industrial assumptions were based on the substantial interference with nature. In all three cases, it affected relations with nearby water reservoirs. A devastated, unnatural coastline making access to water difficult and the lack of beaches in the western part of the city is the reality of the coast in Genoa. A significant contamination of the river Lambro in the Sesto San Giovanni design prevents the use of its waters in the newly proposed park. In Trent, the river Adige had to be separated from the site of investment by a powerful embankment.
4. It is necessary to assess the land contamination caused by years of activity of the factories. This is important for the safety of future residents.
5. The neighbourhood of active industrial areas is responsible for contamination of seawater in Genoa and the river Lambro. In both cases, this fact reduces the attractiveness of the investment. The negative impact has the emission of atmospheric pollution and noise of production processes and transport, which are located nearby.
6. Noteworthy is also the complete lack of natural, high greenery on the plots.

The positive characteristics of these areas include:

1. access to communication provided by the numerous lines of communication means—some of them were previously used for servicing industrial production,
2. location next to the water, which was a prerequisite for the development of the industry. In the case of Genoa and Trent, it assured transportation, and in Sesto San Giovanni, it was used for technological processes of steel,
3. attractive location in the invested urban tissue.

10.1 Genoa, the "Waterfront" Project

Historic Genoa since its inception has been developing based on maritime trade.[4] The city was strongly associated with the sea, the main commercial functions occupied a place near the waterfront and the old port—*Porto Antico*. This concerned both the medieval city and the baroque assumptions at Strada Nuova (Fig. 10.4) (Bonenberg, 2007).

The biggest problems of modern Genoa, limiting its spatial development are topographic situation and location of industrial functions along the western part of its waterfront. Modern four thousand cities develop in a relatively narrow zone of the waterfront, reduced from the land side by steep hills. Access to the sea of a

[4]The first written documents mentioning the Genoese port come from Roman times. On the basis of excavations, it is already known that during the reign of the Etruscans and Greeks, Genoa was a peripheral maritime centre.

Fig. 10.1 New district of Trento. Museum under construction. *Photograph* A. Bonenberg

Fig. 10.2 New district of Trento. Tropical Palm House *Photograph* A. Bonenberg

Fig. 10.3 New district of Trento. *Photograph* A. Bonenberg

densely populated western part of the city is limited by the extensive storage waterfront, warehouses of containers and shipyards. These areas are chaotically built and used selectively. Many elements of the coastal infrastructure from the fifties and sixties were excluded from use. That happened in regions of Pra, Pegli, Cornigliano, Sampierderena.[5] In unfavourable topographical conditions, the residential buildings are climbing up the steep slopes surrounding the city from the north. Multi-storey multi-family buildings pile up in the narrow mountain valleys, frequently with a poor transport accessibility. Many buildings were developed in the period of a nearly wild urbanization of the fifties and sixties. They were built in a very close distance from each other, where the construction site was determined only by land ownership considerations, and not by additional requirements for lighting, access or composition of the whole. The spatial arrangement of Genoa makes the communication between the districts (individual valleys) possible primarily in the coastal strip, which makes the design problem even more complex (Bonenberg, 2011).

[5]Pra, Pegli, Cornigliano, Sampierderena. Districts of Genoa, which coastline in the period from the beginning of the twentieth century till the war times has been developed by expanding shipbuilding, ports and steel mills. Currently, in some places are being made the first attempts to restore access to the sea.

Fig. 10.4 Historic Genoa, the baroque part of the city Strada Nuova, Palazzo Lomellini. *Photograph* A. Bonenberg

The "Genova Waterfront" project[6] was developed in the office Renzo Piano Building Workshop and now it is continued in the form of the initiative Genoa Urban Lab.[7] and proposes to incorporate the coastal zone, covering an area of nearly 250 ha. The idea of the project is the spatial development of the city towards the sea. It assumes enabling access to some parts of the waterfront to residents by reorganization and modernization of the existing industrial functions, which are an important factor in the economic development of the city. In Sestri Ponente port, arrangement of warehouse and industrial functions enabled to locate there the tourist harbour and marina. An important objective of the Genoa Urban Lab is to develop proposals for the transformation of reloading in the district Voltri. In the vicinity of the centre and Porto Antico, the project provides for the opening of a new waterfront between the Old Pier and the fair grounds—*Fiera*. The concept also implies a change in the location of the Genoa airport. According to the proposal of Renzo Piano, it would be located on an artificial island near the port of Sestri Ponente. Among the four above-mentioned strategic elements of the coastal redevelopment, airport location raises the most controversies because of the high cost of implementation. Restoration of the key urban features in the close proximity to the shore recreates the lost communication with the sea, so strongly rooted in the history and culture of Genoa. Project activities seeking to improve the existing situation are concentrated in the coastal area (Fig. 10.5).

The project aims at optimization of the use of the coastal zone, arrangement of the chaotic land development and point annexation of public spaces to the sea. The priority is to activate the waterfront area, while maintaining the industrial functions and the character of the existing building.

The boundary of the revitalized part is designated by the existing range of communication consisting of the railway line and two-lane highways. In the project, it is subject to a partial reconstruction. Between the Old Pier (*Molo Vecchio*) and the area of the Genoese trade fair (*Fiera*), there has been designed a green seaside promenade with commercial functions. The presented concept fully exploits the potential of the sea coast location, and this is a creative continuation of the existing urban layout and opens up better opportunities for the use of the most attractive part of the city.

Many of the proposals of Renzo Piano are in the planning stage, but the only fully realized piece of the revitalization of the coast of Genoa is the project of the Old Port (*Porto Antico*) completed in 1992. It is a good example of activation of the strict city centre—for centuries associated with the maritime transport. Changes in the scale and method of operation of maritime transport in the eighteenth and nineteenth centuries caused that the bustling port—the social space, turned into a closed area: warehousing and industrial, servicing commercial vessels. Since 1903, as the area managed by the Autonomous Port Consortium, was fenced off from the

[6]The waterfront.

[7]Current projects and activities of Genova Urban Lab: http://www.genovaurbanlab.it and developed on the basis of publication: Genova Urban Lab, quaderno 1.

Fig. 10.5 Coast of Genoa: shipyard in the area of Porto Antico. *Photograph* A. Bonenberg

city, blocking direct contact with the sea. The spatial situation of the centre became very unfavourable and already in the interwar period was developed the first plans for the revitalization of Porto Antico. The ultimate excuse for functional and spatial changes became a great five-hundredth anniversary of the discovery of America by Christopher Columbus, celebrated in 1992. Big media event, designed to build a new image of the city required an adequate background. The project entrusted to Renzo Piano included with its range the bay waterfront and the docks plunging into it. He assumed numerous demolitions, adaptations of historic buildings and the construction of modern elements complementing the whole composition. The long-term effect of the investment was the creation of a social communication space, a multifunctional, flexible meeting place as well as for organization of mass media events, concerts and sporting events. Activation followed the introduction of a variety of functions: marina, tourist port, multi-purpose watching arena, conference centre, centre of studies of marine ecosystems, shopping malls, sports facilities, restaurants and hotels. The area regained a representative character. The project required a reference to the main problems:

1. establishment of a relationship with a post-industrial heritage (adaptation of historic buildings),
2. removal of a part of the onerous functions such as warehousing, trans-shipment, leaving the active shipyard at the outlet of the bay,
3. selection and diversification of functions corresponding to the information society: the functions associated with education, access to knowledge, meeting and banquet services, retail, entertainment,
4. providing opportunities to change the purpose and method of use (both for service facilities and multi-purpose arena—creating of flexible spaces),
5. determination of the nature and references of new architectural and urban elements in the Old Port: in order to create a personalized space.

The character of the shaped space thus depends on the relationship between the above factors. The visual consequence of the chosen design philosophy was linking the modern architecture with clear seafaring inspirations with historic features of medieval urban tissue. A rich and highly multithreaded space of social communication was created—in which the spirit of the city past is sensed but also the contemporary aspirations, desires and hopes for the future, prosperous development (Diagram 10.1).

Since its revitalization in 1992, the main functions occupying the waters of the bay of Porto Antico are a yacht marina and a port for tourist boats. Marina is situated in the heart of the bay, and its most prestigious part is the region of "Aquarium" and Magazzini di Cotone (historical buildings of the cotton

Diagram 10.1 Old and new architectural elements in the space of Porto Antico. Historical structures-marked red, modern buildings-marked yellow. (Fig. A. Bonenberg based on www.portoantico.it)

warehouses). The marina is formed along the main pier to which are moored large yachts with lengths from 10 to 55 m. From the main pier diverges smaller ones, with places for small boats. Marina can accommodate a total of 280 boats and yachts, and the main part is located in the western part of the bay. Buildings associated with Marina's facilities include a four-star hotel "Jolly Marina" and many small shops and services. Location inside a circular bay, which is the historical heart of Genoa provides a spectacular entrance in the precincts of the old city to boaters and tourists. When entering the port, on the left side, you can see the aquarium and a glass structure with a tropical forest inside it. On the right side, there is historical Magazzini di Cotone. Nonetheless, before a visitor can see a panoramic view of the harbour, he or she must pass along the active shipyard docks. Leaving the industrial functions in the area of the revitalized port is beneficial because it indicates the nature of the economy of the region: maritime trade and transport. Assessing the form of the marina on the land side, we will see that it is the key element in the city space. Modern solids and a white colour of boats stand out from the colours and textures of the old town, bring into it a new quality.

A significant investment in the Old Port which aim was to revive the tourism has been "Aquarium". In recent years, it has become one of the most visited attractions in Genoa. In addition to the commercial activities, a primary mission of "Aquarium" is to promote the marine environment protection and the responsible management of water resources. This aspect of education is important because of the proximity, omnipresence of the sea in the life of the inhabitants of the city. Contemporary ecological issues presented at the exhibition make people aware of the interrelatedness of life on land and in the sea. With an emphasis on understanding the dangers of natural imbalance and over-exploitation of aquatic ecosystems, the aquarium is based on different methods of generalization of knowledge: workshops, thematic meetings, competitions and exhibitions. It also conducts its own research initiatives funded by the Italian Ministry of Education. The consequence of location of the thriving institution close to the port was a significant increase in a tourist traffic and an economic stimulation of the neighbourhood. As for the marina and the harbour, the presence of "Aquarium" significantly raised the prestige of their location.

Magazzini Cotone is a complex of historic buildings constructed in 1898 by L. Timosci on the artificially created promontory. In the early days of its existence, it was the largest storage space available in the area of Porto Antico. Along with the degradation of the rank of the Old Port area in the twentieth century, progressed the destruction of storage facilities. Adaptations of warehouses were carried out according to plans by Renzo Piano, while maintaining a maximum flexibility of the internal space. The lowest floor of the building was intended for restaurants and bars. Their advantage is the availability from the waterfront and from the open internal passage. Numerous rhythmically repeated openings of storage gates were glazed and in summer evenings they can be opened towards the bay. Bars and restaurants serve the users of the marina and attract many inhabitants. On the upper floors, there is a modern conference centre, with the main auditorium having 740 seats and many smaller classrooms and offices. Another part is the exhibition space with an area of 8500 m^2. It is divided into modules compatible with the internal divisions of the space. Particularly interesting are the parts located on the ground floor because, like bars and restaurants, they enable us to go directly to the promenade running around the bay. In addition to the exhibition centre, in the ground floor of Magazzini Cotone has been localized small services: sailing and diving schools, sailing shops. Like in the case of the aquarium, here the mutual neighbourhood is also beneficial: it attracts a wide range of clients.

The success of the project is the result of the modern architecture form and its relationship with the urban setting. Renzo Piano rightly emphasized the nature of the port, trade and industrial town, taking inspiration from the industrial forms of container ships, tankers, ocean liners. He applies characteristic details, materials and even the way of finishing building elements. His reinterpretation goes so far that the boundary between the modern architecture of the harbour and yachts mooring inside blurs. The result of such a design approach is modern forms alluding to the shapes existing in the environment, other than the architectural elements of the old city. A compositional centre of the Old Port is Bigo, a structure named after the cargo crane for handling of cargo from a ship to the shore. Bigo designed by Renzo Piano is a kind of modern sculpture in the harbour. It consists of steel arms connected with anchor cables, which together with a concrete plinth of the foundation grow from the surface of the water at the edge of the dock. Bigo has the same seafaring style as the other objects in the harbour. It is the spatial dominant feature, the meeting point and purpose of the walks. Over the years, it has become a silhouette promoting the city and the hallmark of Genoa's Porto Antico. The main, longest arm of Bigo, in accordance with the purpose of the original cargo crane, carries a moving structure of the panoramic elevator, from which you can admire the harbour and the old town. To the arms located on the opposite side, there is attached a light steel structure for roofing the multifunctional arena. In winter, there is an ice rink here, and in summer, a roller skating rink or mats for practicing the Oriental martial arts. The roof of the building has a steel structure filled with PTFE material. A white tight material brings to mind the colour and texture of the sail (Diagram 10.2).

The Genoese "Aquarium" is an example of the compositional processing of forms typical for ships. The body of the building is formed linearly, and the aquarium tanks disposed at both sides look like the transport containers. The similarities are also visible in the architectural detail and colours of the structure. The shapes of windows and loggias have a "marine" style: rounded corners and visible rivets of connections with the plane of the wall. Some windows are round and look like traditional bull's eyes. Air-conditioning vents have the shape and colours of the ventilation outlets for passenger transatlantic liners. The facades of the building, in addition to numerous glazing, are dominated by steel wall finish and gray and blue colours. The element of the exhibition of "Aquarium" is *La Sfera*—a spherical greenhouse with a tropical forest inside. Despite its small size, it plays a very important role in the urban interior of the port: thanks to its lightness and the elegance of design; it is very attractive and exposed part of the panorama. The shutters hung to provide protection of greenery from the sun are in the form of

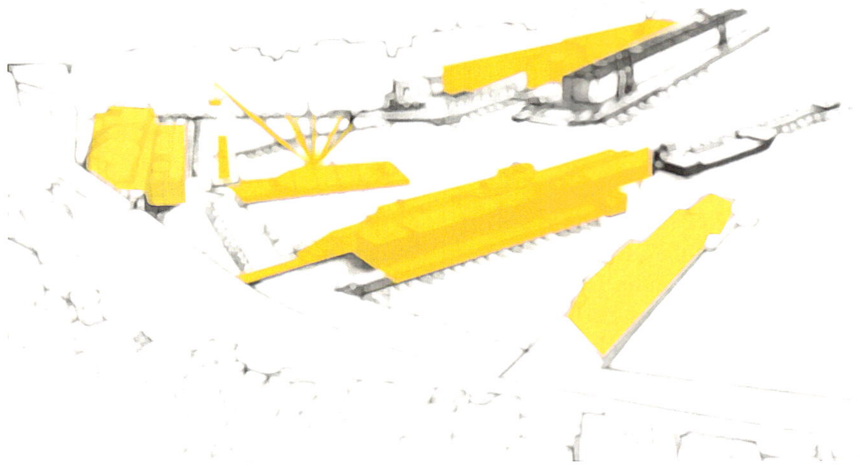

Diagram 10.2 Modern buildings in Porto Antico are marked in yellow. Drawn: A. Bonenberg

triangular sails, visible from the shore of the bay through the transparent walls of the Sphere.

The motivation and philosophy underlying the arrangement of the waterfront was the important role of winds in the sea travels. The artistic installation by Sumusu Shingu makes the air movement to become visually perceptible in the city. The installation consists of tall poles completed with "mobiles" in the form of windmill wings. They move in multiple planes depending on the direction and strength of the wind. "Mobiles" are arranged along the facade of the Magazzini Generali building alternating with rows of palm trees and have similar proportions to them. These two types of elements: animate and inanimate ones form a "grove" of mobile and dynamic objects. The construction of wind turbines is maintained in the stylistic convention similar to the "Aquarium", Bigo and urban details of the port. White canvas of triangular wings is a parallel of tight sails. The presence of a regular grid of trees and windmills introduced an element of geometry and sense of order into the space. Elements of greenery and the art of the leading artist made the square to obtain the rank of elegant "living room" of the city—a traditional interior with an innovative, avant-garde furniture (Figs. 10.6, 10.7 and 10.8).

All newly constructed buildings within Porto Antico are similar in style: they follow, imitate the form of yachts and sailboats that are a regular element of the landscape. The obtained result is a stylistically coherent space with high aesthetic quality, which does not distort the readability of the urban tissue. The realization of the project resulted in the imposition of a new concept on the old one, which shall be without prejudice to the readability of historical systems.

Fig. 10.6 Pool Porto Antico in Genoa—a view from the sea. *Photograph* A. Bonenberg

Creation of the customized space based on historical urban tissue

The issue of juxtaposing modern architecture with historic urban tissue is often the subject of heated debates in the society. The revitalization project concerning the Old Port was a major challenge in this respect, because the development area coincides with the area of the creation of Genoa. Architectural themes of the medieval town are here well legible in the walls of the existing buildings. Thus, the history of the region Porto Antico is in fact the history of the same Genoa: here was located the main commercial port, a source of wealth and power of the residents. The area loaded with centuries of history forced the designer to pursue a coherent strategy in relation to the existing cultural heritage. In the project for Porto Antico, Renzo Piano follows the below principle:

1. existing historic buildings receive a new function after renovation. Modern elements of buildings are located in such a way as not to disturb the perception of solids of monuments,
2. newly established buildings belong to a completely different style than the historical ones—to the stylistics of ships and boats that are a permanent, natural and inseparable part of the city space. With this solution, the most recent structures fully refer to shapes existing in the environment, although the forms are not derived from the brick architecture.

Fig. 10.7 Dock overlooking the sphere—a part of the exhibition in the Genoa's aquarium accommodating a tropical forest. *Photograph* A. Bonenberg

In urban terms, the design of Renzo Piano is not entirely subordinated to the historic tissue of the city, but it is well adapted to it, respecting and highlighting the strengths of the existing space, such as observation planes, compositional closings and spatial dominant features (Diagram 10.3).

Due to choosing the path of dialogue with the cultural layer, traces of the ancient way of life, buildings have not been defaced. New concepts imposed on the old ones without causing distortion to the image of ancient traces. Highlighting of the

Fig. 10.8 Mobile sculptures by Sumusu Shingu. *Photograph* A. Bonenberg

historical heritage promotes the role defined by the designers, that is of the first contact with the city.

The project of Porto Antico by Renzo Piano was a huge success, because it restored to city residents a valuable public space, arranging and reorganizing the functions of the port. The success of the project lies in its multilayer structure, richness of meaning, diversification of the functional space. The creation of the customized space based on the historical urban tissue.

Diagram 10.3 Historical buildings in Porto Antico are marked in red. Drawn A. Bonenberg

10.2 Milan, the "FALCK" Project

The subject of the project for Milan is an area belonging in the past to a steelworks "Falck". The steelworks unit was founded in 1906, in the town of Sesto San Giovanni near Milan. For a hundred years, it was the largest private manufacturer of steel and iron in Italy. The factory area was known as the "Italian Stalingrad" because of the concentration of heavy industry and supporting branches. Industrial functions were laid out on the area of 150 ha.

In the seventies, the profile of business operations of the group changed. At the post-industrial areas remained many interesting architectural structures, which are a memorial and a symbol of the glorious days of "Falck"—the company inscribed in the history of Milan. The historic metallurgical halls have a high potential of creating the individual quality of space.

Visitors will be able to feel the large scale of establishment of the old factories. The existing industrial buildings are the background and point of reference for the new architecture, clearly contrasted and cutting off from the old metallurgical facilities. New forms are light and transparent, clearly delimiting overlapping points of two different systems: the old one, which is a remnant of the old ironworks and the modern one, carrying the aesthetic and social message of the twentieth-first century. The industrial nature of the buildings gives to the place a unique character. From the first sketches of Renzo Piano, it was clearly visible that some of the objects remaining after the steelworks must become an integral part of the new project. The architect summarized his concept as follows: "In my project, I will follow in the footsteps left by the factory "Falck". The area of "Falck" is the perfect site for greenery, I dream about the park that would be so great as to make your head spin" (Pastore 2005). According to this concept was created a modern urban composition, based on the idea of a large enclave of greenery within the city.

In the new district was proposed a rich functional program, tailored to the needs of areas where development is associated with the creative industry. Plans of adapting historic buildings for functions associated with the mass media—new headquarters of Rai television and university facilities associated with culture and science are associated with the creation of a society based on a permanent access to information. It is important here to create inspiring spaces that encourage to creativity.

There is planned the construction of a new Museum of Science and Technology and a new stadium of the football team Inter Milan. There are designed extensive systems of public space for 13,240 new residents, whose number could increase to 30,000 over the time.

An interesting aspect of the project is the Sesto San Giovanni Park, which is an important element of the urban design. The local press called it the "Milan Central Park", since it will cover the area of 45 ha. The concept envisages the introduction of greenery on areas where it has been absent for a hundred years (Giorgetta 2006). Within the park, there will be some high-rise buildings, with the construction suspended on 12 m above the ground. The intensive, high greenery of plantings is be used at several ways. Firstly, it is to assist in creating a favourable microclimate in the residential and working areas. Secondly, the plant material produced by such a large park is to be processed as biomass for energy used for the housing settlement. The research and experimental element is development of a theoretical model of a settlement that would be self-sustaining in terms of energy. The proposed water system is divided into outdoor pools and channels, whose task is to maintain the proper humidity conducive to plant growing in a very dry area, and the system of underground installations allowing to obtain the heat from deep layers of the ground. The development area of the project has ground water at high temperatures, which is an essential element of the energy strategy of the new district.

The river Lambro, running along the site boundary, is visually integrated into the composition; however, due to the high level of industrial pollution, it cannot be used more widely. The system of modern water basins is an important element of the urban composition. The project emphasizes the importance of social, economic and environmental factors.

10.3 Trento, Project for the Post-industrial Areas of Michelin

Historical Trento is a coherent urban organism. In 1927, outside the historic centre of the city was established a tire factory Michelin. The area was convenient for the industry, situated on the bank of the river Adige. Nearby was led the railway allowing the transport of raw materials and products. Back in the eighties, the factory employed thousands of people, being a major employer in the region. After the closure of production in the nineties, the municipal authorities decided to allocate the plot for the construction of a new district of the city. The area is closed: it is delimited from the north side by the active railroad tracks and the road lane and from the south side—by the river. This situation results in spatial and functional cutting off of the area from the neighbouring parts of the city. Although the plot is located close to the historic centre, the railway line, road and river are important spatial barriers.

In 2000, the area occupied by the factory Michelin became the object of a new design by Renzo Piano. The urban structure proposed by the Master shall function in the space independently: the adjacent residential buildings do not have specific aesthetic values so the designed system creates a new compositional quality within itself, and it creates personalized sequences of the urban interiors. New public buildings assume the role of the spatial dominant features, organizing the designed urban interiors. The form of the residential buildings defines the boundaries of public and semi-public spaces. Facades of buildings bring a distinctive rhythm and articulation. One can discern here the inspiration in the surrounding nature—forests and hills around the city. This is also reflected in the applied materials: wood and stone, in the dynamic lines of roofs formed in the shape of alpine peaks visible in the distance. The design concept for Trent implies the desire for associations to nature, not a continuation of existing patterns of urban development. The park, which is an integral part of the establishment, is situated along the Adige River and is a part of a wider regional concept of ecological corridors along the riverside areas and flood banks. Vegetation of the park emphasizes the urban layout of the settlement forming walls of urban interiors, framing views of the river, setting landscape axes.

The river is of high landscaping importance in the mountain valley. Her presence shapes the microclimate, greenery and nature. Unfortunately, for security reasons, its course is interfered with high flood blanks that cause a functional and scenic separation from the area realization of the project. The main channel which is the interpretation of the Adige meander introduces a unique atmosphere in buildings of the district. Shallow pools designed as a floor around the building of the museum in the north and the conference centre in the south form the background for a favourable exposure of the bodies of both buildings. Reflection, multiplication of the modern design will give lightness to the whole composition.

The settlement received an extensive functional program: the central part is formed by residential and office buildings, and from both sides, it is surrounded by cultural and commercial facilities. The centre focusing the life of the district is a natural history museum with a large communal library and a media library. The body of the museum closes a composition of settlements from the north. On the south side are located a concert hall and conference spaces with hotel facilities.

Architecture of the designed office buildings concentrated along the railway line allows the arrangement and distribution of interior design in many ways. Offices, conference rooms are divided by glass walls with an integrated office furniture system. The obtained effect, combining of transparent elements with the built ones gives the impression of unity of an office; it visually expands its scale—

simultaneously allowing to achieve the acoustic comfort. Office spaces in this project are varied and offer a very good sun exposure.

In the building of the Museum of Science, the generally available spaces of social communication include the main foyer on the ground floor and a media library.

10.4 Building an Image Based on the Relationship with the Existing Context of the City and Development of the Social Communication Space

Despite differences in scale and location of the discussed projects, we can distinguish the principles that guided Renzo Piano in creating customized spatial solutions. They refer to the classical principles formulated by the prominent theorists of urban planning: Christopher Alexander, Louis Kahn, Kevin Lynch. In the book "Image of the City", Lynch (1997) is writing that the fundamental, positive feature of the city is its integrity that is a feeling that the urban body is a harmonious whole. The spatial integrity is a prerequisite for the creation of social communication sites. It is also an important feature of multithreading—characteristic for the information society.

And so, despite the ongoing spatial transformation, vanishing of certain functions, intensification and expansion of the building structures, despite the construction of flexible, multifunctional spaces, the basic issue is to create a sense of unity of the new establishment with the existing urban tissue. At the level of composition, it can be achieved by extending the relevant communication, functional and visual links. An example of the viewing axis of the newly designed street, culminating in the historic spatial dominant feature commonly associated with Trento (view of the Palazzo delle Albere), creates a personalized space by linking it to the historical architecture. In urban projects of Renzo Piano, the important unifying element tend to be public green areas, common recreational spaces, continued in the urban or even regional system. An integrating role of the public recreation greenery, the natural environment cannot be overestimated.

Despite many similarities, the analysis of solutions shows that the projects represent three different relationships existing between the newly proposed or revitalized area of the city and its existing parts. These relationships consist of:

1. continuation of the existing system (Genoa),
2. creative "imposition" of the new system on the existing one (Milan),
3. creation of a completely new part in the vicinity of the existing one (Trento) with preservation of the basic visual relationships with the elements of a historical tissue.

The choice of the method used by Renzo Piano depends on broad environmental and location conditions, including the nature, culture and tradition associated with the contemporary needs of the functional and technical requirements (Diagram 10.4).

In the first two cases (Genoa, Milan), an incorporation of new parts of the city in the old tissue is carried out according to the traditional principles. The new system,

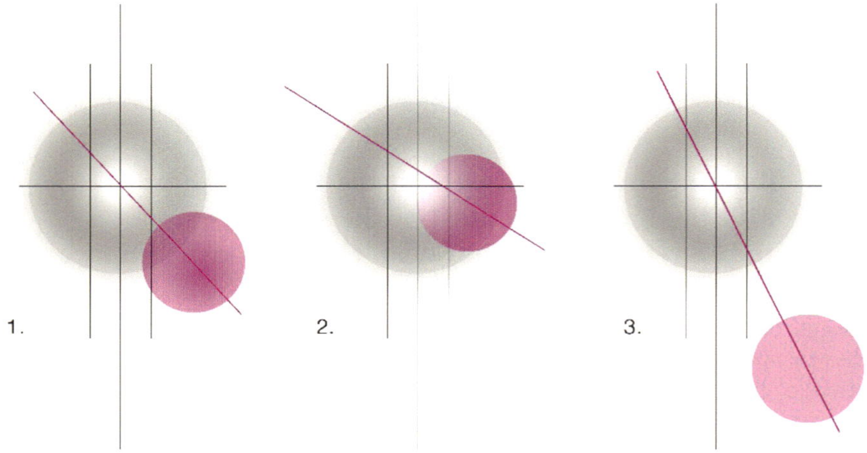

Diagram 10.4 Schemes illustrating the three types of relationships that can exist between the newly designed area of the city and its existing parts. 1. Scheme No. 1—Continuation of the existing system (Genoa). 2. Scheme No. 2—"Imposition" of a new system on the existing one (Milan). 3. Scheme No. 3—The new part is formed in the vicinity of the existing one (Trento)

coexisting with the old one, uses its landscape dominant features and formally valid points. In this case, the role of the new elements of the urban structure is a harmonious continuation and enrichment of existing assumptions.

In the case of an independent composition (Trento), a new urban system creates independently the most of the important points of reference: dominant features, formally valid points, public, semi-public and private zoning, and the only historical reference point is the above-mentioned Palazzo delle Albere.[8]

References

Giorgetta, F. (2006). RPBW Fabrica di idee nell'ex area FALCK Sesto san Giovanni, Milano. *Abitare, 465*, 126.

Lynch, K. (1997). *The image of the City*. London: The MIT Press.

Pastore, L. (2005). Piano: Sesto sarà la città del sapere. *La Repubblica, 16*.

Bonenberg, A. (2007). Analiza jakości i kompozycji funkcji w projekcie rewitalizacji portowej bramy miasta na przykładzie Porto Antico w Genui [w] Odnowa krajobrazu miejskiego–Bramy i granice miasta, monografia, Materiały konferencyjne międzynarodowej konferencji naukowej wydziału Architektury Politechniki Śląskiej, Gliwice. 285.

Bonenberg, A. (2011). Water Systems for New Living Quarters in Projects by Renzo Piano [w] Water in the Townscape, monograph, A. Januchta-Szostak, Poznań: Wydawnictwo Politechniki Poznańskiej.

[8]Palazzo delle Albere is a sixteenth-century villa—a fortress of the prince bishops Madruzzo. Currently, the Museum of Modern and Contemporary Art of Trento and Rovereto (Museo di arte moderna e contemporanea di Trento e Rovereto).

Chapter 11
Conclusion

A set of statements concerning the media, information and communication technologies and cityscapes included in the book aims to create a possibly complete picture illustrating the multithreaded nature of the issue. The essence of a scientific achievement described in this publication is based on the formulation and analysis of the research problem assuming the relationship between the mass media transmission, information society and spatial behaviour of people. Spatial behaviours affect the transformation of the environment, cities and architecture.

The research problem was analysed in the context of the cultural, social and technical conditions. The adopted research method is consistent with the assumptions of research by design—a study based on examples. The direction of the spatial transformation becomes a medium revealing cultural, economic and social relations. The emergence of a social values based on access to information shapes and changes spatial behaviours.

The availability of information and communication technologies affects the directions of transformation of space and is characterized by:

1. Possibility to use information and communication technologies in the management of resources, better and more complete use of the architectural spaces. Constant communication via the Internet is a tool by which it is possible to make an appointment of a group of people or to sequence the use of the space.
2. Use of flexible solutions in architecture: to facilitate the adaptation to dynamic changes in the method of use. Use of mobile elements (decomposition) and the negotiable spaces. The condition of a spatial flexibility is particularly important when optimizing management through information and communication technologies.
3. Building a space supporting creativity and innovation activities by improving the quality of the environment, degree of individualization, opening to the cultural and natural context of the architectural solutions.
4. Building sites integrating a community in the real space, not just virtual. Focus on neighbourhood activities—in the era of remote work, the "era of freelancers",

A. Bonenberg, *Cityscape in the Era of Information and Communication Technologies*, The Urban Book Series,
https://doi.org/10.1007/978-3-319-69542-6_11

it is the quality of interaction with others and the quality of the space that determines where people settle—the distance from the place of employment becomes a secondary factor.
5. A positive perception and evaluation of diversity, multithreading in both the cultural and spatial sphere of the city.

The applied research method is based on the analysis of case studies that is the projects of urban districts by Renzo Piano created in cooperation with the author. The representativeness of the examples is guaranteed by the prestige of the architectural design Renzo Piano and numerous gained awards (including a prestigious Pritzker Award). The "Cityscape in the Era of Information and Communication Technologies" orders, therefore, the transformation of architectural space in dynamic conditions of re-evaluations in the global economy based on creativity and innovation activities—as the basic elements of development.

GPSR Compliance

The European Union's (EU) General Product Safety Regulation (GPSR) is a set of rules that requires consumer products to be safe and our obligations to ensure this.

If you have any concerns about our products, you can contact us on ProductSafety@springernature.com

In case Publisher is established outside the EU, the EU authorized representative is:

Springer Nature Customer Service Center GmbH
Europaplatz 3
69115 Heidelberg, Germany

Batch number: 10845468

Printed by Printforce, the Netherlands